Field Experience Guide
Resources for Teachers of Elementary and Middle School Mathematics

for

Van de Walle

Elementary and Middle School Mathematics

Fifth Edition

prepared by

Jamar Pickreign
State University of New York, Fredonia

PEARSON

A and B

Boston   New York   San Francisco
Mexico City   Montreal   Toronto   London   Madrid   Munich   Paris
Hong Kong   Singapore   Tokyo   Cape Town   Sydney

ISBN 0-205-39453-1

Printed in the United States of America

10 9 8 7 6 5 4 3   08 07 06 05

Contents

PART TWO: RESOURCES FOR TEACHING

"Purposeful mental engagement or reflective thought about the ideas we want students to develop is the single most important key to effective teaching. Without actively thinking about the important concepts of the lesson, learning will not happen."

- John A. Van de Walle

To make such learning possible, teachers need to create mathematical environments, pose worthwhile mathematical tasks, use cooperative learning, use a variety of models as thinking tools, encourage discourse and writing, require student justification of responses, and listen actively. This guide is intended to help the prospective teacher develop understanding of and gain some experience with these important strategies.

The guide is divided into two parts:

Part One: Observing and Assessing in the School, presents useful information to introduce the reader to the elementary and middle school mathematics classroom and what being an effective teacher of mathematics involves. Readers are encouraged to visit classrooms to gain a sense of the mathematics environment and learning expectations, and to learn about and reflect on the teacher's role in developing effective mathematics teaching strategies. There are numerous forms, guides, and recommendations to assist the reader in learning about and reflecting on these important strategies.

Part Two: Resources for Teaching, presents nearly thirty lesson ideas as well as over 55 reproducible black line masters. There are also three complete assessment activities with scoring rubrics, as well as several sets of instructions and ideas for creating your own manipulative models.

A unique feature of this guide is the approach taken in presenting the lesson ideas. Each idea is presented in a problem-based lesson format with an accompanying activity sheet using the problem-based lesson planning strategy outlined in Chapter 6 of "Elementary and Middle School Mathematics: Teaching Developmentally" by John A. Van de Walle. However, only the first lesson presented is complete. The reader is encouraged to reflect on their learning about effective mathematics teaching and to edit, revise, or otherwise adapt any of the lesson ideas to his or her needs. Each lesson also contains "Links" to "Key Teaching Strategies," that is, the effective teaching strategies. These "links" are comments on how the lesson idea may be connected to effective teaching strategies. Again, in only the first lesson idea, are these links complete. In the other lesson plans, readers are encouraged to reflect on the key teaching strategies and enter their own comments on how the lesson idea may connect to these strategies. In this way, these lesson "plans" are not intended to be used as printed, but are intended to encourage the user to think about how they can use these ideas to bring about "purposeful mental engagement" with the mathematics.

ACKNOWLEDGEMENTS

Special recognition goes to the following people who reviewed early drafts of the Guide:

> Kathleen Chamberlain, Ph.D., Lycoming College
> Sylvia Taube, Sam Houston State University
> Janie Cates, State University of West Georgia

Much appreciation and gratitude go to Tom, Traci, John, and most especially, Jocelyn, Hannah, Liam, and Kelley.

Part One

Observing and Assessing In The School

1. Getting Started

This section provides some useful forms as you begin your field experiences. Section 2 includes more in-depth observation and other worksheets to help you assess, reflect and improve on your experiences.

The first few times you visit a school during your field experience can be a little overwhelming. Use the following handouts to help you assess and reflect on what you see and learn:

- **Field Experience Cover Sheet**
- **First Impressions of a Classroom**
- **Field Experience Activity Log**
- **Simple Lesson Observation**

When you conduct any field experience assignment, attach the following cover sheet and give to your instructor.

Details

Your name: _____   Course: _____

School visited: _____

Date: _____

Time spent at the school: _____   Grade: _____

Host teacher: _____   Number of students: _____

Type of experience

_____Observation

_____ Interview

_____ Teaching

_____ Other: _____

As you enter a classroom, imagine you are a prospective parent visiting it for the first time trying to decide if it's the right environment for your child. As you walk around, jot down your responses to these questions:

1) Are there any mathematics pictures or posters on the walls? What do they depict?

2) Are there any bulletin boards with mathematics information or learners mathematics work?

3) How are the learners' desks or tables arranged? Why do you think they are arranged that way?

4) Are there any math manipulatives evident? Are they accessible to the learners?

5) Are there any computers in the class? Is it connected to the Internet? Is it used for mathematics?

6) What are the learning expectations for this class? How do you know?

Use this log as a handwritten journal to record the dates, times and short descriptions of your experiences. The first table gives an example.

Date	Times	Activity
3/6/03	8:30 – 9:00	Observed start of class
	9:00 – 10:00	Helped set up counting activity
3/7/03	1:00 – 2:00	Observed Mrs. Smith give presentation on fractions
	2:00 – 2:30	Interviewed Mrs. Smith

Date	Times	Activity

Use this handout to help you observe your first lesson. After the lesson has finished think about what surprised you about the lesson. What was effective and what was not effective in the lesson? Think about what students had learned by the end of the lesson.

1) How does the teacher greet the students at the start of the class?

2) What kind of lesson is it? What are the main learning goals? How do you know?

3) How does the teacher introduce the lesson?

4) How does the teacher communicate with the class? In what ways do the students communicate with each other?

5) What strategies does the teacher use to get students attention and to keep them on task?

6) What strategies does the teacher use to assess students understanding?

7) What resources, books, or materials does the teacher use?

8) Are any special arrangements made for the students with physical or learning disabilities?

2. Looking Closer:
Becoming an Effective Teacher

Effective teaching strategies are meant to promote, as John Van de Walle suggests (Van de Walle, p. 32), "purposeful mental engagement or reflective thought about the ideas we want students to develop" which he indicates is the "single most important key to effective teaching."

This section is organized around the following teaching strategies that you should think about if you want to become an effective teacher:

- Creating an Effective Mathematical Environment
- Posing Worthwhile Mathematical Tasks
- Using Cooperative Learning Groups
- Using Models as Thinking Tools
- Encouraging Student Discourse and Listening Actively

The observation and interview worksheets in this section will help you to assess the classes you visit during your field experiences, and will help you think about ways to carry out these teaching strategies when you teach.

Creating an Effective Mathematical Environment

Effective mathematics teaching depends on the kind of mathematical environment that is established in a classroom. The *Principles and Standards for School Mathematics* suggests that this environment is affected by the decisions teachers make, the conversations teachers orchestrate, as well as the physical setting they create. As the *Principles and Standards for School Mathematics* states (NCTM, 2000, p. 18):

"Teachers establish and nurture an environment conducive to learning mathematics through the decisions they make, the conversations they orchestrate, and the physical setting they create. Teachers' actions are what encourage students to think, question, solve problems, and discuss their ideas, strategies, and solutions. The teacher is responsible for creating an intellectual environment where serious mathematical thinking is the norm. More than just a physical setting with desks, bulletin boards, and posters, the classroom environment communicates subtle messages about what is valued in learning and doing mathematics. Are students' discussion and collaboration encouraged? Are students expected to justify their thinking? If students are to learn to make conjectures, experiment with various approaches to solving problems, construct mathematical arguments and respond to others' arguments, then creating an environment that fosters these kinds of activities is essential."

This section contains the following observational and interviewing activities to help you assess and reflect on the Mathematical Environment in your school.

- **Observing the Mathematical Environment**
- **Sketching a "Bird's–Eye View" of a Mathematics Classroom**
- **Interviewing a Student about Attitudes toward Mathematics**
- **Observing Students with Special Needs**
- **Interviewing a Teacher about Creating a Good Environment**
- **Creating an Effective Mathematics Environment – Some Guidelines**

Observing the Mathematical Environment

Observe a mathematics class and as you watch think about whether the teacher has created an effective environment for learning (see Van de Walle, p. 36). Write down your answers to the following questions:

1) Did the teacher seem to encourage mathematical thinking? If so, what were some of the teacher's actions that seemed to promote thinking?

_____

_____

_____

_____

2) Did the learners make conjectures or engage in mathematical arguments? Were they expected to defend, or support their arguments and conjectures? How do you know? Describe the teacher's actions that facilitated this.

_____

_____

_____

3) Circle any of the verbs below that you think describe the activities students are asked to do during the lesson.

explore	investigate	conjecture	solve	justify
represent	formulate	discover	construct	verify
explain	predict	develop	describe	use

4) Recall the way the room was set up. Sketch the classroom. Where were the learners? How were they seated? Were there any mathematical models (manipulatives) available? Were they accessible? Were there any computers. Were they used? How? Based on the physical setting of the classroom, how do you think the teacher views mathematics teaching and learning?

_____

_____

_____

_____

Sketching a "Bird's-Eye View" of a Mathematics Classroom

Teaching and learning can be greatly affected by the physical arrangement of a classroom. Observe the classroom you are in and draw a sketch of the location of the desks, whiteboard (or chalkboard), resources and teacher. Think about how the arrangement helps or hinders the learning of mathematics. How would you arrange the classroom to better help student learning?

Interviewing a Student about Attitudes toward Mathematics

Ask a student about his/her perspective on his/her math class. As you talk to the student, think about how the mathematical environment shapes his/her attitudes.

Try the following attitudinal questions derived from the National Assessment of Educational Progress (NAEP) fourth grade mathematics test. Each question can be followed with a "Why?"

1) Do you think everybody can do well in math if they try?

2) Are you good at mathematics?

3) Do you like mathematics?

4) Is math more for boys?

5) Is math mostly memorization?

6) Is math useful for everyday problems?

7) Is there only one correct way to solve a math problem?

8) Do people use math in their jobs?

9) If you had a choice, would you continue to go to math class?

10) Do you understand what goes on during math?

Observing Students with Special Needs

An effective mathematical environment includes all learners. A good teacher should pay careful attention to a child with special needs and how he or she learns and design instruction (not content) that maximizes the strengths of that student (see Van de Walle, chapter 7). Observe a class where a student identified with special needs is included. Watch and note your answers to the following questions. Think about the effectiveness of the support for students with special needs.

1) What categories of disability or exceptionality does the students have?

2) What kinds of resources are available in the classroom/school to help the student?

3) How did the teacher modify or adapt his/her instruction to the student?

4) Who provided help for the student during the class? How much time was spent helping the student?

5) Did the student successfully complete the activities in the lesson? Why or why not?

Interviewing a Teacher about Creating a Good Environment

Ask a practicing teacher the following questions. What does his/her response reveal about his/her beliefs about the role of environment in learning mathematics? (Questions derived from *Professional Standards for Teaching Mathematics*, NCTM, 1991, p. 57)

1) How do you decide how much time to provide learners to explore sound math and grapple with significant math problems?

2) How does the way you've arranged the room (e.g., the seating, the location of materials, etc.) affect the learners' learning of mathematics?

3) I understand we want learners to develop mathematical skills and proficiency. I've heard that providing appropriate contexts encourages this. How does one do that?

4) What do you do about the learner who says that he/she doesn't like math or is afraid of math?

5) Do learners ask questions in math or make conjectures? Do you encourage this? How?

6) I've learned in class that math should make sense. What do your learners do that sends the message that they are making sense of math? How do you encourage that?

Creating an Effective Mathematics Environment–Some Guidelines

As you think about creating an effective mathematics environment, it is important to remember that it involves more than just the physical arrangement of the room and its material. Much of an effective environment is established by the expectations for dealing with mathematics learning that a teacher communicates both explicitly and implicitly to learners.

"Good mathematics is NOT how many answers you know... but how you behave when you don't know."

-Author unknown

A teacher establishing an effective learning environment for mathematics recognizes this and encourages behaviors and actions of learners that promote mathematical understanding.

The following are some suggestions derived from the *Professional Standards for Teaching Mathematics* (NCTM, 1991) to help guide you in creating an effective environment.

Provide sufficient time for learners to explore significant mathematics and to grapple with important ideas and problems. Consult your schools curriculum documents to aid you in identifying those important ideas.

Consider the ways children think and learn mathematics (see Van de Walle Chapter 3) and make use of appropriate models. Use the physical space available in your classroom in ways that facilitate understanding of mathematics. For example, consider having children act out significant problems or become "living" mathematical models.

Think about the value and utility of the mathematics that you want the learners to learn. Share with your learners this value and utility. Where in the world might they encounter such mathematics? Make their learning relevant whenever possible.

Become as knowledgeable of the mathematics content and effective ways of *learning* that content as you can. This way you can better consider every child's response and respect the variety in thinking that can be witnessed in a classroom. Be aware that there is no requirement that children must like mathematics, but that your job may be to help them to develop an appreciation for mathematics.

Encourage all learners to work independently or collaboratively as needed to make sense of mathematics. Encourage learners to ask questions, to make and share mathematics conjectures, and have them provide reasons to support their conjectures. Provide a "risk-free" environment that encourages meaningful "grappling" with significant mathematics.

Posing Worthwhile Mathematical Tasks

(Please check Part II Section 3 for a wide selection of tasks and activities you can use in your teaching)

Another key element of effectively teaching mathematics is the use of worthwhile mathematical tasks or activities. Well-selected tasks have the potential to pique curiosity and, thereby draw learners into mathematics. Good tasks are based on sound and significant mathematics and provide opportunity to engage in problem solving, to communicate about mathematics, and to see ways that mathematics is connected to other aspects of the learners' experience.

The *Professional Standards for Teaching Mathematics* (NCTM, 1991) discusses *Worthwhile Mathematical Tasks*. The following guidelines are based on this discussion. When planning a mathematics lesson that is task-based, these guidelines offer descriptions of, characteristics of, and selection criteria for worthwhile mathematical tasks.

I.   **The selection (or design) and implementation of Worthwhile Mathematical Tasks are teachers' responsibility.**

II.   **Worthwhile Mathematical Tasks can be selected from a variety of sources.**
    A.   Published Material
    B.   Teacher-made Material
    C.   Capitalizing on a "Classroom Moment"

III. **Worthwhile Mathematical Tasks are characterized by the following:**
    A.   They connect Mathematical Thinking with the Mathematical Concept
    B.   They capture the curiosity of the learner
    C.   They can be approached from a variety of interesting and legitimate ways
    D.   Some worthwhile tasks are open ended
    E.   They facilitate Significant Classroom discourse by providing students with opportunities to:
        1.   Observe
        2.   Compare and Classify
        3.   Generalize
        4.   Analyze and Synthesize
        5.   Hypothesize and Predict
        6.   Evaluate

IV.   **As a teacher selecting a task, you should consider the following in determining whether the task is a *worthwhile* mathematical task:**
    A.   The *Mathematical Content* of the task is of utmost importance
        1.   Does the task represent the concepts and procedures appropriately?
        2.   Does the task involve the students in "doing" mathematics?
        3.   Does the task foster relevant skill development?
    B.   What you know about your specific students must be considered as well
        1.   Does the task "fit" with what you know about your students?
            a.   Consider psychological, cultural, sociological, and political perspectives

2.   Does the task "fit" with what you know about your students academically?
        a.   Consider:
            i.   Their Prior Knowledge
            ii.   Their Strengths and Weaknesses
            iii.   Their capacity for Intellectual Challenges
3.   Does the task appeal to your students interests, disposition and experiences?

C.   Finally, what you know about *How* students learn mathematics is necessary to assure the worthiness of a task.
     1.   Examples:
        a.   Developmental Approaches to learning.
        b.   Constructivism
     2.   Some sources of information on *How* students learn mathematics.
        a.   Research
        b.   Experience
            i.   Actual teaching
            ii.   Observation of teacher
        c.   Tasks that provide insight on student thinking.

This section contains the following observational and interviewing activities to help you assess and reflect on the tasks used in lessons you observe. A number of worthwhile mathematical tasks are included at the back of this Guide.

- **Observing an Activity in Class**
- **Planning a Problem-Based Lesson or Activity– Guidelines**
- **Teaching via Problem-Solving–Personal Reflection**
- **Assessing and Evaluating Students' Understanding– Rubrics**
- **Observing and Interviewing Students to Inform Instructions**

John Van de Walle suggests some specific questions one should ask regarding the use of classroom activities (Van de Walle, p. 51). While observing a class, note when the teacher starts to prepare students for an activity. As you watch the activity develop, write down your responses to these questions. Think also about whether the task fits the criteria for a Worthwhile Mathematical Task outlined earlier.

Observation:	Notes:
1. How is the activity done? How do the children do the activity or solve the problem? What materials are needed? What are the steps students need to take? What is written or recorded by the student?	
2. What is the purpose? What mathematical ideas will the activity develop? Are they concepts or procedures? What kind of mathematical connections will there be? Can students extend ideas?	
3. What does the teacher do? How does the teacher set up the activity (demonstrations, instructions, examples, etc.)? What does the teacher do when the students are on task? How does the teacher deal with student problems? What does the teacher do once the task is complete?	
4. Did the activity accomplish its purpose? Was the activity successful? How do you know? What evidence of learning did the learners present?	

Planning a Problem-Based Lesson or Activity – Guidelines

John Van de Walle, offers a nine step procedure to help you plan a problem-based activity or lesson. Please refer to Van de Walle, pp. 81 – 83 for the specific descriptions of each step.

Step 1: Begin with the math!

Step 2: Think about your students.

Step 3: Decide on a task.

Step 4: Predict what will happen.

Step 5: Articulate students' responsibilities.

Step 6: Plan the BEFORE portion of your lesson.

Step 7: Think about the DURING portion of the lesson.

Step 8: Think about the AFTER portion of the lesson.

Step 9: Write your lesson plan.

The following is a possible format that you might use to plan a problem – based lesson. This is the one used for the lesson plans in Part Two of this field guide

Title:	Optional title of lesson
Grade:	Write the grade level for which the lesson is intended
Math:	Here write in <u>simple words for you</u> exactly what you want the children to learn.
Task:	Describe the problem task.
Expectations:	Describe expectations. Groups, pairs, or individuals? List transparencies, manipulatives, handouts, etc.
BEFORE activity:	Describe the activity.
TIME allotment:	Before _____ During _____ After _____
NOTES (as needed):	Possible hints. Key ideas for discussion. How you plan to conduct discussion (*before* and *after*). Learners or groups from whom to gather data. What to do with written work.

A Note on Planning a Problem – Based Lesson

Depending upon when in your program sequence you learn to write lesson plans, the steps for planning a problem-based lesson may present challenges. You may have noticed the "lesson plan" of an experienced teacher very seldom resembles the "lesson plan" format you may have learned in an education course. For example, the typical "objectives, rationale, materials, set induction, procedures, evaluation" style lesson plans that can often be multiple pages may cause you to question why you should be expected to produce such a plan, when you know "real" teachers just "write a couple sentences in their plan book." You should be aware that attention to articulating each of the typical lesson plan characteristics is important for developing the mind-set for planning lessons. You will realize that "real" teachers have thought about each of those characteristics and have planned them out, but only need a "couple of sentences in their plan book." This "covert planning" is advocated in the "Planning a Problem-Based Lesson" steps.

For example, step 1 encourages you to identify the mathematics for the lesson. Many teacher candidates comfortable with the "objective, etc." format attempt to equate this step with stating an objective. Objectives are dense statements that encompass content, method, and evaluation. Thus, teacher candidates may not focus on just the mathematics. You need to clearly identify the mathematics you plan to attend to.

Also, recognizing that not all performance type activities need to be assessed and evaluated with a rubric will help you understand the importance of the after portion of the plan. The after portion is indeed part of the steps, but you should articulate it clearly in the Notes portion of the plan.

It should be noted, too, that not all lessons in mathematics need be "situational." Many activities that can be used in a problem – based lesson will not be. A game called "Salute" in which three participants, two of whom each hold up a number card (seeing each others but not their own), and the third person calling the sum of the number cards for the others to determine their own card is certainly not situational, but is still a problem-based activity. You need to develop the ability to identify an adequate problem.

Many beginning teachers and teacher candidates design lessons that involve teaching by telling. Identifying worthwhile problem situations or activities can prove challenging. As you examine the sample lessons presented in this guide, pay careful attention to the identification of the math and the problem and the description of the before and after activities.

Teaching Via Problem-Solving- Personal Reflections

Having observed a number of lessons, think about how you would respond to each of these "frequently asked questions" about teaching via problem solving. Compare your responses to those provided in Van de Walle (pp. 53 – 54)

1) How can I teach all the basic skills I have to teach?

_____

_____

_____

2) Why is it okay for students to "tell" or "explain" but not for me?

_____

_____

_____

3) Where can I find the time to cover everything?

_____

_____

_____

4) Is there any place for drill and practice?

_____

_____

_____

5) What do I do when a task bombs?

_____

_____

Assessing Students' Understanding through Analytic Rubrics
(Grades 3–5)

A well-prepared rubric (see Van de Walle p. 64) provides you the opportunity to consider what information is important about your students understanding. It helps you to define how you *know* whether your students understand. Ask your supervising teacher if you can set up the following geometry problem to a small group of students:

Describe the following picture. Write down what you see.

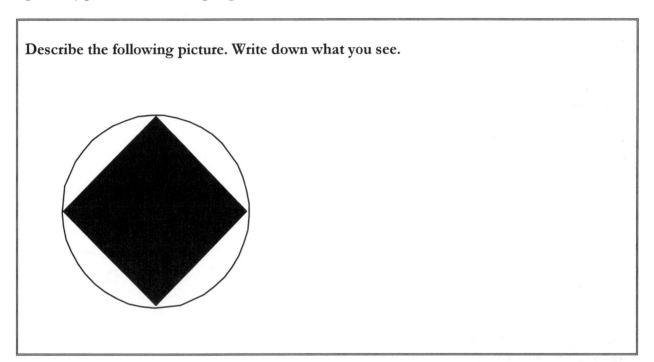

Collect the students' work and use the following analytic rubric to assess their understanding. How well did the students perform? How accurate do you think the rubric was in assessing student understanding?

RUBRIC CHARACTERISTIC:	POINTS:
A white circle	1
A black square	1
All vertices of square touch circle	1
Square is "standing: on one vertex. Accept "looks like a diamond"	1
TOTAL POINTS	4

Alternatively, use the holistic rubric on the next page during completion of the problem.

Assessing Students' Understanding through Holistic Rubrics
(Grades 3–5)

The following is a template for an *observational* holistic rubric which you could use to assess student performance while they are doing the problem from the previous worksheet. See Van de Walle (p. 68) to see a completed example of this type of rubric.

The Math: _____

SUPER **Clear understanding.** **Communicates concept in multiple representations.** **Shows evidence of using idea without prompting.** Content-specific Descriptors	
ON TARGET **Understands or is developing well. Uses designated models.** Content-specific Descriptors:	
NOT YET **Some confusion or misunderstands. Only models idea with help.** Content-specific Descriptors:	

Analytical Rubrics

Questions you should consider when creating analytical rubrics might include, but are not limited to:

1. **Might one characteristic be "worth more" than another? Why?**

2. **Have I included all "crucial" characteristics? Could these characteristics be completely adhered to and yield an "incorrect" solution?**

3. **How many of the "crucial" characteristics should the student get to be "acceptable?"**

4. **What is "unacceptable?"**

Holistic Rubrics

Below is a procedure for holistic rubrics. After having children work on a problem and recording their responses...

1) Do a "first pass." Identify the responses as "got it" or "not yet." Sort the student responses into these two piles.

2) Ask yourself, "What is it about the responses in the 'got it' pile that demonstrates that the students understand the math in the problem?" Ask yourself, "What is it about the responses in the 'not yet' pile that demonstrates that the students are struggling with something in the problem?" Write these down as characteristics of "got it" and "not yet."

3) Now sort through each pile creating two new piles for each of the "got it" and "not yet" piles. In the "got it" pile, look for evidence of "full accomplishment" of the problem. What is that evidence? Write it down as a characteristic of that performance. Keep in mind that you should look for whether the response effectively communicates a solution, not how long it is. Look also for evidence of "substantial accomplishment." These are going to be responses of students who "get it," but may not be judged at the same level of effectiveness as the others. Record the evidence as characteristics of that performance.

Sorting through the "not yet" pile may be more of a challenge. You are judging how much struggling with the major concepts took place to determine which responses are those in need of some instruction versus those in need of significant instruction. Look for evidence of these struggles and record these as characteristics of these performances.

4) Organize these characteristics into a 4-level rubric. See page 105 for a sample task and rubric.

The following are some questions to think about when observing students learning. The goal is not to teach, but to get information to inform instructional planning.

1) Does the student use materials purposefully? Is it directed toward solving the problem or question presented? If not, how could you redirect the student?

2) Does the student demonstrate evidence of understanding the problem or question presented? If not, could you rephrase the problem or question? If that doesn't work, how does that inform your planning? Was the problem beyond the student's capability? What kinds of "prerequisite" understanding are necessary and could you approach the development of that understanding before beginning the problem?

3) Does the student demonstrate evidence of having several strategies with which to approach and/or complete the problem or question presented? Assuming the problem "can be approached from a variety of interesting and legitimate ways," (p. 16) how can you help the student discover some other approaches?

4) What vocabulary does the student use to describe the problem or question or materials? Does it include precise mathematics vocabulary? Can you plan for effective vocabulary development to address any deficiencies?

5) Does the child ask questions? At what cognitive level is the question he/she asks? How can you facilitate the development of higher-level thinking questions?

6) How persistent in trying different strategies is the student? Did the student become frustrated? Why or why not? How can you help the student develop good problem solving habits? (See Van de Walle Chapter 4)

The goal of an interview is to find out where a child is at a particular time in terms of concepts and procedures (Van de Walle, p.76). The goal is not to teach but to get information to inform instructional planning. Ask your supervising teacher if you can interview a student and then follow the steps below.

MATH

Rational Counting.

TASK

Provide learner with a random number of counters that is more than twenty. Ask the learner, "How many counters do you have?"

KEY QUESTIONS

(Note that some of these questions are not necessarily questions that you will ask aloud, but questions you ask yourself in gathering information).

a) What strategy (if any) was employed? (e.g., count by ones, twos, etc.,)
b) Were the counters moved into pile(s) as they were counted? How?
c) Were any counters counted more than once? How many?
d) What number language are they using to count?

Depending upon the observations made with regard to the above questions, you might consider asking aloud the following questions:

e) "You counted by [ones, twos, fives, etc.], now count them by [ones, twos, fives]." Observe whether the learner can count on from a number using a different counting factor (e.g., there are 23 counters, counting by fives gets to 20, can the learner count on from 20 by ones the remaining three?).
f) If the counters were placed into piles, say, "Tell me why you chose to do that."
g) If any counters were counted more than once, present the same counting task with fewer than ten counters. Ask yourself whether the child conserves number (i.e., the child can avoid "double counting" despite the arrangement of the counters, and knows when to stop counting to have counted them all). The task may be developmentally beyond the learner right now.

As John Van de Walle (Van de Walle, p. 33) suggests, putting students in groups of three or four to work on a problem is an extremely useful strategy for encouraging the discourse and interaction envisioned in a mathematical community. Other advantages include the opportunity to learn about and respect each other's diversity, and to learn how to support each other's learning.

- **Composing Cooperative Groups – Some Guidelines**
- **Observing and Assessing Group Work**
- **Assessing and Evaluating Group Work – Some Guidelines**

The most common cooperative groups are small learning groups. These are no more than six learners in heterogeneous groups usually assigned to complete a particular task. The following are very simple guidelines drawn from literature (Aronson, 1978; Johnson & Johnson, 1975; Kagen, 1985; Sharan & Sharan, 1976) for putting this type of group together.

1) Only two to six learners per group

2) Assign learners heterogeneously along a desired domain (e.g., cognitive ability, learning style, disposition, etc.) Most often this is done by cognitive ability, but teacher judgment is crucial.

3) Depending on the nature of the task, assign specific "jobs" to each learner in the group (e.g., recorder, reporter, facilitator, discussant, etc.), but remind the learners that they are each responsible for actively participating and learning.

4) Provide clear and specific expectations for the group interactions.

5) Arrange the room to facilitate group activity.

Observing and Assessing Group Work

Observe a lesson that features group work and write down your responses to the following questions. Think:

1) How does the teacher organize the groups? How are they seated? What is the composition of the groups?

2) Does the teacher go over rules for group work? What are they?

3) How long does it take for students to get into groups?

Choose one group; sit close and listen. Note your answers to the following questions:

1) How well does the group interact? Does everyone contribute equally? Does the group stay on task?

2) Do any students have difficulty participating? How does the teacher, assigned aid, other students, or you help that student?

3) How do the groups record and report their answers?

4) How does the teacher assess the group's work?

Assessing and Evaluating Group Work – Some Guidelines

In assessing group work, you must clearly identify what it is you want the learners to gain from the work. If it is only the correct solution to a problem, you'll likely find yourself confronting parental fears that their children are carrying other children in the group task. Thus, there must be more to be gained for the learners in using groups. Among these include the development of communication skills, development of a sense of responsibility, and the development of a positive disposition and ability to work with others toward a goal.

The following are considerations for creating a rubric for assessing and evaluating group work.

1) Development of Content Knowledge. Ensure that each learner has learned the math identified in your plan. Although learners may rely on other group member to solve the problem for them, you should assess and evaluate whether all group members grasp the math content. This presents opportunities for other forms of assessment. How will you *know* that the learners understand the content?

2) Development of Communication Skills. Students should be assessed on how well they communicate within the group and out of the group. Devise criteria for what is acceptable evidence of each groups ability to communicate with each other and with other groups and with the teacher.

3) Development of a Sense of Responsibility. How well does each group member carry out his or her assigned responsibility? Devise ways to identify evidence that each job is appropriately attended to. How will you *know* that the "observer" has fulfilled his or her responsibility?

4) Development of a Positive Disposition and Ability to Work with Others. All members should participate. Consider how well the group members ask questions and involve silent members. Consider how well the group members share ideas and opinions and how well they learn to think as a group, rather than as individuals.

Models help children explore ideas and make sense of them. Manipulatives, calculators and computers should be readily available for student use as a regular part of your classroom environment.

There are many types of models that can be used to represent mathematical concepts. Concrete models, or *manipulatives*, are frequently used to initiate understanding in mathematics. As "thinker tools," it is important to remember that the mathematics is not in the manipulative. It is in the relationships that the learner constructs through use of representational models. Use of manipulative material does not result in mathematics concept acquisition unless learners actively construct these relationships. Learners, consequently, can be "on task" with the manipulative material but "off task" with the mathematics intended.

There are numerous commercially manufactured manipulatives available. Among them include base blocks, connecting blocks, centimeter rods, various counters (colored tiles, two-sided counters, etc.), attribute blocks, pattern blocks, fraction sets, geoboards, rulers, scales, thermometers, measuring cups, play money, tangrams, pentominos, geoblocks, dice, spinners, dominos, and calculators.

Many manipulative models, however, can be made by teachers, as well as learners, using commonly available material. Having learners create their own manipulatives may provide additional benefits such as a sense of ownership in the model and, depending upon the directions given for creating them, opportunity to gain conceptual insight.

In addition to more traditional "hands-on" concrete models, there has been a surge in availability of "virtual" models or "on-line thinker tools." Many of these make use of current technologies allowing users to manipulate computer generated graphics similarly to traditional models. There are a number of these models available on the internet.

Some of the best can be found through NCTM's web site *Illuminations* (http://illuminations.nctm.org) or through their *e-Examples* (http://standards.nctm.org/document/eexamples/index.htm).

Other sources of these "applets" (or "mini-applications") include the *National Library of Virtual Manipulatives* located on-line from Utah State University at http://matti.usu.edu/nlvm/nav/index.html.

As with any model, care must be exercised to select those that will facilitate thinking and learning. A good rule of thumb is to decide whether a model allows the user to think, or if it does the thinking for her or him. (See Van de Walle, chapter 8 for reviews).

- **Surveying Classroom Manipulatives**
- **Assessing Manipulatives on the Web**
- **Observing Manipulatives Use in the Classroom**
- **Using Manipulatives– Some Guidelines**

Surveying Classroom Manipulatives

This sheet allows you to survey the kinds of manipulatives available to help children learn mathematics in a classroom or school. Investigate the availability and use of these resources by talking to teachers, supervisors and students. As you do your research, observe which mathematical topics the manipulatives are used for, their condition and whether they have been made by the students.

Manipulative	Available?	Topic(s) used for?	Condition? (good/fair/poor)	Made by students?
Base Blocks				
Connecting Blocks				
Centimeter Rods				
Various Counters (Colored Tiles, Two-Sided Counters, Etc.)				
Attribute Blocks				
Pattern Blocks				
Specific Fraction Sets				
Geoboards				
Rulers				
Scales				
Thermometers				
Measuring Cups				
Play Money				
Tangrams				
Pentominos				
Geoblocks				
Dice				
Spinners				
Dominos				
Calculators				
Other:				

Choose three of the applets of "web-based" manipulatives from the web sites listed at the end of Van de Walle Chapter 8. For each applet, apply the selection criteria provided on pages 110 – 111 of chapter 8. How do these applets "measure up?"

Applet Assessment Guide

Applet source: http://_____   Recommended Grade level:_____

1) What does this applet do better than conventional methods?

2) How are students engaged with the *content*?

3) How easy is it to use?

4) Are there instructions?

5) Describe the conceptual information provided.

6) If there is a drill associated with the applet, how are wrong answers handled? Are they handled in pedagogically appropriate ways? How do you know?

7) Are there provisions for record keeping for the teacher? Describe them.

8) Is there printable material available? Is it "worthwhile" material?

Observing Manipulatives Use in the Classroom

Observe a class where a teacher is using manipulatives to teach mathematics. As you watch, write down your answers to the following questions. Think about the benefits and problems of using manipulatives.

1) What manipulatives are being used? What mathematical concept is being modeled?

2) How are the manipulatives stored? In the classroom?

3) Does the teacher demonstrate how the manipulatives can be used to model mathematics concepts?

4) How are the manipulatives distributed to students?

5) Does the teacher allow the learners time to explore and familiarize themselves with the manipulatives?

6) Do the students remain on task, or are they distracted by the manipulative?

7) What procedures are used to collect the manipulatives at the end of the lesson?

Using Manipulatives- Some Guidelines

Successful use of manipulatives can be positively affected by the way they are introduced and incorporated into the classroom. Joyner (1990) offers some simple "rules." As you plan lessons that include the use of manipulatives, think about the following:

1) <u>Packaging the manipulative</u>. The models to be used should be packaged in such a way as to allow ease of distribution at the start and collection when finished. Re-sealable plastic bags, or boxes can be used to store enough of the model for an individual learner to use. Besides the obvious "classroom management" advantages, it encourages learners to be responsible for their tools.

2) <u>Free exploration or play</u>. Often the introduction of a model results in a level of "distraction." Don't try to jump right to the mathematical modeling. Allow the learners some time to explore and familiarize themselves with the material. The comfort they gain with the material, as well as the diminishing novelty of the material, can reduce the "distractibility" when using them to model mathematics concepts.

3) <u>Demonstrate</u>. Show the learners how the material can be used to model mathematics concepts. Remember that the mathematics is not in the material; a bean is a bean, but a bean can be a model of the number one.

Encouraging Discourse, Writing, and Listening Actively

John Van de Walle suggests that the more we try to explain something, the more connections we will search for and use in our explanations: "talking gets the talker involved." (Van de Walle, p. 33) When students are asked to respond or critique others they are forced to really think reflectively about what is being said. Similarly, writing can help structure students' thoughts and rehearse explanations of problems in their minds. Encouraging students to explain why, to tell how, and to detail their ideas makes them aware that mathematics is not mysterious or unfathomable.

"Because the discourse of the mathematics class reflects messages about what it means to know mathematics, what makes something true or reasonable, and what doing mathematics entails, it is central to both *what* students learn about mathematics as well as *how* they learn it" (NCTM, 1991, p. 54).

Active listening requires the teacher to believe in a student's ideas. When a student responds to a question, the teacher should respond in a way that encourages the student to elaborate: "Tell me more about that, Sally," or "I see. Why do you think that Tim?"

Use the following handouts and guidelines to help you assess and reflect on the discourse interactions you observe.

- **Observing Student-Teacher Interactions**
- **Observing Student-Student Interactions**
- **Encouraging and Assessing Student Responses– Some Guidelines**
- **Assessing Writing Activities**
- **Looking for Active Listening**

Observing Student–Teacher Interactions

Classroom discourse includes the ways of representing, thinking, talking, agreeing and disagreeing in the classroom. It is the way in which ideas are exchanged and what the ideas entail. The teacher must present a model of this. Observe the communicative interactions in a classroom and respond to the following questions derived from the *Professional Standards for Teaching Mathematics* (NCTM, 1991).

1. Does the teacher do most of the talking or is there a "collaborative" atmosphere in the class where the teacher speaks and students listen, then the students speak and the teacher listens?

2. What evidence do you see that the teacher is listening carefully to student ideas?

3. Does the teacher pose questions and tasks that elicit, engage, and challenge each student's thinking? How do you know?

4. What evidence do you see that the teacher is interested in the learners reasoning through issues and ideas and presenting evidence of their understanding?

5. Do students initiate problems or questions?

6. Do students make conjectures and offer arguments to support them?

7. Does the teacher allow students to struggle with problems? For how long?

8. What does the teacher do to get learners to clarify and justify their ideas?

9. Are students encouraged to present solutions?

10. Is the teacher willing to pursue an interesting idea raised by the class discussions or student question? How do you know?

11. How does the teacher encourage class participation?

12. Which of these sequences tends to occur–
    a. the introduction of mathematical symbolism and notation followed by a discussion of what it means, or
    b. a discussion of a concept followed by the association to mathematical symbolism and notation?

13. Who is the mathematical authority in this classroom– the learners, the teacher, or the textbook?

Meaningful interactions among students are an indicator of an effective mathematics environment. Take some time to observe the interactions among students in a mathematics classroom.

Provide your responses to the following questions derived from the *Professional Standards for Teaching Mathematics* (NCTM, 1991):

1. **Do students try to convince themselves and one another of the validity of particular representations, solutions, conjectures, and answers? What sorts of things do they do?**

2. **Do students listen to, respond to, and question one another? Describe the tone of these interactions.**

3. **Do students initiate problems and questions among each other? What is the nature of these problems and questions?**

Encouraging and Assessing Student Responses– Some Guidelines

Teachers can encourage discourse in a number of ways. The *Professional Standards for School Mathematics* (NCTM, 1991) identified a number of "tools for enhancing discourse." The use of

> " • computers, calculators, and other technology;
> • concrete materials used as models;
> • pictures, diagrams, tables, and graphs;
> • invented and conventional terms and symbols;
> • metaphors, analogies, and stories;
> • written hypotheses, explanations, and arguments;
> • oral presentations and dramatizations" (NCTM, 1991, p. 52)

can provide vehicles for engaging students in the interactions that make up classroom discourse.

Another important idea in the Effective Mathematics classroom is getting the child to verify and justify his or her mathematics thinking and understanding. Remember also that listening *is* assessing and requires good questioning strategies. We ask questions for a variety of purposes: motivation, challenge, provoke student response, get students to evaluate, focus on process, guide, diagnose, review or summarize, evaluate, communicate interest, encourage exploration, promote reflection, to name a few. What is perhaps as crucial as the questions we ask is how we listen to student responses and what we communicate to learners as a result. The following offer some suggestions and guidelines for active listening and questioning strategies derived from *Geometry and Visualization* (Hoffer, 1977).

1. Remember to provide sufficient wait time (3-5 seconds).

2. Avoid answering your own questions. Also avoid repeating your questions. Repeating the question may breed inattention to the first statement of a question. If you get nervous when no student volunteers right away, a little more wait-time or a rephrasing of the question may get results.

3. Don't be quick to "correct" the student response. Especially avoid interrupting a student as he or she is responding. Let the student know that you value his or her contribution.

4. Request validation and justification for all responses (not just the incorrect ones). Without the expectation that all answers will require justification, students will assume their response is "wrong" if a teacher asks, "How did you get that?"

5. Avoid too many questions which can be answered with "yes" or "no," "fill-in-the-blank" questions like "The answer to an addition problem is called …?" and pseudo-questions like "This is a fraction, right, Tom?"

6. Think about the "tone" of your question. What is the covert message? Are you expecting too much? Too little? Are you devaluing learners? Be careful to avoid criticism or attacks, e.g., "Why didn't you do your assignment?"

(Continued…)

7. Think about creating an environment that promotes student questions.

8. When learners do ask questions, take time to be sure you understand what they are asking.

9. Encourage participation in discourse with prompts like: (*adapted from Geometry and Visualization*)
    - "Give me another example."
    - "Do you agree with that, Hannah?"
    - "Will that work with _____, as well?"
    - "How do you know?"
    - "Why do you disagree with Liam?"
    - "Jocelyn, can you convince the majority that they are wrong?"
    - "Will that always work?"
    - "Say that another way."

10. Avoid repeating a student's response. If you think the rest of the class did not hear the response, ask the learner to either speak more loudly or try turning more to face the class more fully. This has the benefit of promoting more student-student interactions.

"Oral communication, valuable as it may be, is gone at the end of the day" (Van de Walle, p. 69).

Recall that we don't always need to evaluate student work. Sometimes we just want to collect information to inform our teaching. Writing activities are an important way to provide this information. Many different opportunities for writing exist in the classroom. Van de Walle offers suggestions for writing activities for all students.

Through examining student writing, teachers learn about their students' understanding of important mathematics ideas. The following list of questions can serve as a guide to examining students' written work. Responses to these questions for each learner should be saved somewhere to help you in shaping his or her learning.

1. **What evidence does the learner present that he/she understands the concept about which he/she is writing?**

2. **If in a problem solving situation the learner is asked to write how he/she knows his/her solution is correct, does he/she provide a rationale or a description of what they did? (If it is a description, what can you as a teacher do to shape future responses into rationales?)**

3. **What kind of conceptual errors does he/she make? What evidence is there for this in the writing? What will this tell you?**

4. **What thoughts do they express that show interest in the concept that deserve comment? (Van de Walle suggests that to not respond to such things sends the message that you as a teacher do not value their responses.)**

Hearing is the registering of audible sound. Listening, on the other hand is the active engagement in determining the meaning of the sounds. Promoting reflective thinking is a major goal of effective mathematics education. Listening to student responses, observations, or conjectures is a way to facilitate reflective thinking.

A teacher listens to the learners to gain understanding of what they are saying and insight into what they may be thinking. John Van de Walle suggests, "active listening requires that we believe in children's ideas" (Van de Walle, p. 33). Learners will sense that we value their thoughts and ideas, and will be inclined to think more and develop understanding.

As you observe classroom discourse, provide your response to the following questions.

1. **When listening to a student, does the teacher appear to be present? That is, do you observe that the teacher is focused on what the learner is saying? How do you know?**

2. **Does the teacher attempt immediately to evaluate the learner's response, observation, or conjecture?**

3. **Do you perceive the teacher as being open-minded? What behaviors of the teacher tell you this? (Watch for behaviors that might suggest whether the teacher is thinking about the content what the learner is saying rather than his or her delivery).**

4. **Does the teacher ask questions to help clarify his or her understanding of the learners' response? Write down some of these questions.**

5. **Does the teacher provide feedback that indicates he or she is listening and is genuinely interested in what the learner is saying? Write down some examples of what the teacher offers as feedback (e.g., maintains eye contact, nods, says "um-hmm," etc.).**

Aronson, E. (1978). *The jigsaw classroom.* Beverly Hills: Sage Publications.

Johnson, D., &Johnson, R. (1975). *Learning together and alone.* Englewood Cliffs, NJ: Prentice-Hall.

Kagan, S. (1985). *Cooperative learning: Resources for teachers.* Riverside, CA: University of Califomia, School of Education.

Sharan, S., & Sharan, Y. (1976). *Small group teaching.* Englewood Cliffs, NJ: Educational Technology.

Hoffer, Alan R. (1977).*Geometry and Visualization* Mathematics Resource Project. Sunnyvale, CA: Creative Publications.

National Council of Teachers of Mathematics (1991). *Professional Standards for Teaching Mathematics.* Reston, VA: The Council.

National Council of Teachers of Mathematics (2000). *Principles and Standards for School Mathematics.* Reston, VA: The Council

National Center for Educational Statistics(2003). *National Assessment of Educational Progress* http://nces.ed.gov/nationsreportcard/

Van de Walle (2004) Elementary and Middle School Mathematics: Teaching Developmentally, 5/e

Joyner, J. M. (1990). "Using manipulatives successfully" *Arithmetic Teacher* 38 (2), 6 - 7.

Part Two
Resources for Teaching

3. Worthwhile Tasks and Activities with Lesson Plans

In this section, a number of lesson ideas and activities are presented. Each is intended to be a starting point for you in planning for your particular classroom, and presents an opportunity for reflection, and response to the ideas presented in part one of this field guide. You as the teacher should consider the relevant steps of the 9-step problem-based lesson planning format. Think about how such plans incorporate the effective teaching strategies, and edit or modify the plans accordingly.

To aid you, a lesson plan template based on the 9-step format is offered. Within this template are specific questions you might find helpful in tailoring these ideas to fit your situation. In addition, the template includes questions in a side-bar that can help you in thinking about how the plan incorporates or depends on the effective teaching strategies.

The first activity and lesson is a complete plan with full commentary as a means to illustrate the purpose of this section. The responses to these questions are by no means intended to be the "final word." They are to stimulate reflection on the strategies, ideas, and concepts expressed both here and in the Van de Walle textbook.

For each of the remaining ideas or activities, some of these questions have been answered leaving you the opportunity to complete the plans, tailoring or modifying them to fit your specific needs. Classrooms of today are much more diverse. Learners can differ in skills, rates of learning, language proficiency, abilities to abstract, physical abilities, sociability, motivation, as well as prior knowledge. As you consider each lesson idea, think about ways in which you could differentiate the task to meet the demands of diverse learners.

Consult the Van de Walle textbook, chapter 9 "Planning in the Problem – Based Classroom" for additional support.

Topic	Activity	K-2	3-5	6-8
Developing Early Number Concepts and Number Sense	1. Pipes!	X		
	7. The Find!	X		
	15. Odd or Even	X	X	
	26. Up to 57	X	X	
Developing Meanings for the Operations	6. Factor Quest		X	
Helping Children Master the Basic Facts	25. Target Number	X	X	X
Whole-Number Place-Value Development	22. Rename That Number!	X	X	
Strategies for Whole-Number Computation	17. Cakes, Cakes, and More Cakes	X		
Computational Estimation with Whole Numbers	29. Magic Age Rings		X	X
Developing Fraction Concepts	19. Squeeze		X	
Computation with Fractions	5. Fraction Pictures		X	X
	18. Life with Fractions		X	X
Decimal and Percent Concepts and Decimal Computation	2. Compensation Decision		X	X
Developing Concepts of Ratio and Proportion	12. Interference		X	X
	20. Illustrating Ratios		X	X
	23. Fruit Punch!		X	X
	30. Grocery Store		X	X
Developing Measurement Concepts	4. Garden Fence		X	
	8. Building Bridges		X	
	11. Cover All		X	
	16. Hiking		X	X
	21. Right Angle Challenge		X	X
	28. Bolts and Nuts!		X	X
Geometric Thinking and Geometric Concepts	10. Stacking Blocks	X	X	
	13. Triangle Communication		X	
	24. Tangram Polygons		X	
Exploring Concepts of Data Analysis and Probability	3. Candy Store		X	
	9. Ranking Theme Parks		X	X
	27. Mean Basketball!	X	X	
Algebraic Reasoning	2. Compensation Decision		X	X
	8. Building Bridges		X	
	14. Tricking the Trickster	X	X	X
Exploring Functions	8. Building Bridges		X	
Developing Concepts of Exponents, Integers, and Real Numbers	25. Target Number	X	X	X

Lesson Template

Grade: For what grade is this intended? Can it be modified for other grades?

Math: What is the math? Write it in the form, "I want the learners to…"

Task: Describe the task and write as a problem.

Expectation: What are the expectations for performing this task? List what you expect the learners to produce and how you expect them to do it.

BEFORE activity: Specifically describe how you plan to get the learners "ready to learn the math."

Time allotment: Before ≈ minutes **During** ≈ minutes **After** ≈ minutes.

Notes

possible hints: (When considering whether your learners have the necessary background for the ideas in this activity, are there any hints that you could offer that makes the content attainable?)

key ideas for discussion: (In addition to the identification of "the math," what other ideas are important to discuss?)

conducting discussion:

Before: (Write some questions that you could ask to stimulate discussion and get the learners ready to address the activity.)

After: (Think about the important concepts that you wanted the learners to learn. Write some questions that can stimulate discussion to tell you whether the learners "got it.")

specific groups or children from whom to gather data: (As part of the 9-step process, you need to think about your learners. The diverse classroom that is likely yours, presents the need to assess various levels. Are there any specific groups that you need to look closely at?)

what to do with written work: (Is there to be written work? What specific information do you plan to collect from the written work? Why do you want to collect that information? This last question is critical in helping you to decide what you will do with that work. For example, if the work presents evidence that learners understand a particular concept, then you may decide to include that evidence in a learners portfolio. Or you may decide to reflect on that information to help you in planning follow-up lesson/activities. Whatever the goal, have a specific plan.)

Linking to Key Teaching Strategies

Effective Mathematics Environment
How will this activity be affected by the environment? How can this activity facilitate the development of a conducive environment?

Worthwhile Mathematical Task
Consider the characteristics of worthwhile mathematical tasks. What characteristics does this task exhibit? How can it be modified?

Cooperative Learning
What opportunities are there in this activity for cooperative learning? How would you organize the groups? How would you assess their learning?

Models
What different models could be used in this activity? How could you facilitate the connection of symbols and concepts with the models?

Discourse, Writing and Active Listening
What types of writing could be incorporated in this activity? How does the activity facilitate discourse?

What questions could you pose to encourage learners to justify themselves in this activity? Remember that active listening is a way to foster an environment that is characterized by students presenting mathematical arguments.

1. Pipes!

Grade: PreK-2 (can be modified to address what you know about your students).

Math: Part-part-whole relationships. I want the children to learn that ten can be renamed in different ways as two parts.

Task: Mr. Jones is replacing some pipe in his house. He needs 10 feet of pipe but doesn't have a single 10 foot pipe. He does have several shorter pieces. He has pipe that is 1 ft, 2 ft, 3 ft, 4 ft, 5 ft, 6 ft, 7 ft and 8 ft in length. What should he do to get 10 feet?

Expectation: Have children in groups of three or four. Give to each group a set of Cuisenaire Rods consisting of 1 white, 1 red, 1 lt green, 3 purple, 1 yellow, 1 dk green, 1 black, 1 brown, and 1 blue. Have them construct the pipe in parts and record all the possibilities for the needed 10 ft that they can find. When done, discuss the different solutions they found. Provide activity sheets for various "pipe" lengths. (See attached)

BEFORE activity: Present the task orally and brainstorm possibilities with the whole class. Be prepared for some children to propose going to the store and buying a new pipe. Some children may question how pieces of pipe can be fitted together. Some may not see that the pieces can be fitted together because they are only thinking in terms of the whole.

Time allotment: Before ≈ 5-10 minutes **During** ≈ 20 minutes **After** ≈ 20 minutes.

Notes

possible hints:
"Rather than purchase a new pipe, how could we use what we have?" "Is there some way to connect short pipes to make another longer pipe?"

key ideas for discussion:
Shorter pipes connected to make longer pipes, i.e., part-part-whole relationship

conducting discussion:
before: Accept all brainstorming suggestions, listen actively (ask for elaboration of students' responses; nod and encourage others to contribute to discussion, e.g., "what does that make you think of, Jenny?"
after: Have children present their solutions and explain why they work, have other children listen to explanations and challenge them to verify each others solutions.

specific groups or children from whom to gather data:
none

what to do with written work:
Information desired is whether learners are able to rename numbers as two parts as illustrated by the completed activity sheet. Include the activity sheet in learner portfolios.

Linking to Key Teaching Strategies

Effective Mathematics Environment
By posing the task initially you have opportunity to establish a problem-solving atmosphere. The "before" discussion has potential to demonstrate that mathematics can be thought about and possible approaches discussed before any "action" takes place. The teacher's position as one who is asking rather than telling promotes this kind of environment.

Worthwhile Mathematical Task
The context, if elaborated with such things as "My shower sprung a leak spraying water all over my house" or "I'm learning about plumbing from my uncle/friend" etc. puts the problem into a meaningful context that may pique the children's interest. This problem, because of the part-part-whole relationship has many legitimate approaches to its solution which can lead to interesting classroom discussions about the different approaches as well as the patterns and relationship the various pipe lengths have to a total length.

Cooperative Learning
This task lends itself to cooperative learning processes. The task recommends groups of three to four children to work on the problems. It is reasonable, depending on student needs, to organize them as a cooperative group with specific assigned tasks such as recorder, presenter (for class discussion), and leader. Holding each child accountable for understanding may encourage the group to learn from each other.

Models
Making centimeter rods available as models allows children to re-represent the verbal problem in concrete ways. Paper recording sheets allow them to re-represent the concrete models as pictorial models.

Discourse, Writing and Active Listening
Teachers can encourage discourse and/or writing through well asked questions. Examples:

What should we do if we need to replace a pipe that is X ft long, but don't have one that long?
How can we rename X?

These can provide stimulus for either class discussion or journal writing (depending upon the writing ability of the learners)

Active listening is a way to foster an environment that is characterized by students presenting mathematical arguments. Child: "You can only use two fives. Two of anything else is either too long or too short" Teacher: "I see. Can you show me what you mean?" Child can demonstrate. Teacher: "What do you think of Michael's explanation, Isobel?"

Pipes!

For each pipe, write the colors used and a number sentence for those colors:

```
┌─────────────────────────────┐
│                             │
│                             │
└─────────────────────────────┘
```
__dark green and purple__   6 + 4 = 10

```
┌──────────────────────────┐
│                          │
│                          │
└──────────────────────────┘
```
_____

```
┌──────────────────────┐
│                      │
│                      │
└──────────────────────┘
```
_____

```
┌────────────────────┐
│                    │
│                    │
└────────────────────┘
```
_____

```
┌─────────────────┐
│                 │
│                 │
└─────────────────┘
```
_____

Find different ways to create each of these Pipes using 2 Cuisenaire® Rods!

```
┌──────────────┐
│              │
│              │
└──────────────┘
```
_____

```
┌──────────┐
│          │
│          │
└──────────┘
```
_____

2. Compensation Decision

Grade: 5-6 (can be modified to address what you know about your students).

Math: I want the learners to compare arithmetic and geometric growth functions. The task depends on skills of adding decimals and estimation.

Task: Jocelyn wanted to make some extra money. Her father offered to pay her for odd jobs around the home for a week and gave her a choice of two options. The first option was that her father would pay her $1.25 for the week. The second option was that her father would pay her in the following manner for a week: on Monday he would give his daughter $0.01; on Tuesday $0.02; on Wednesday $0.04 and so on through Sunday. What would you tell Jocelyn to do, so she can earn the greatest amount?

Expectation:
Learners work in groups of three or four.
Learners engage in cooperative work to solve the problem.
Learners should use tables, charts, or graphs to support their solution.

BEFORE activity: Review addition of decimals (problem can be modified to avoid decimals)

Time allotment: Before ≈ 5-10 minutes **During** ≈ 15-20 minutes **After** ≈ 20 minutes.

Notes

possible hints:
Provide students with real or fake pennies or coins to use as manipulatives

key ideas for discussion:

conducting discussion:
before:

after:
1. Have groups present how they solved their problem.
2. Have students explain which way Jocelyn would earn more money.
3. Suppose these options extended for a month, what would happen?

specific groups or children from whom to gather data:

what to do with written work:

Compensation Decision   Name: _____

Jocelyn wanted to make some extra money. Her father offered to pay her for odd jobs around the home for a week and gave her a choice of two options. The first option was that her father would pay her $1.25 for the week. The second option was that her father would pay her in the following manner for a week: on Monday he would give his daughter $0.01; on Tuesday $0.02; on Wednesday $0.04 and so on through Sunday. What would you tell Jocelyn to do, so she can earn the greatest amount?

Use this space to solve this problem. You can draw pictures or write number sentences to help you. Be sure to explain how you know that your solution is the one that will earn Jocelyn the greatest amount!

3. Candy Store

Grade: 5-6 (can be modified to address what you know about your students).

Math: I want the learners to analyze data and statistics and use them to make decisions and predictions.

Task: You and a friend decide to open a candy store. What is involved in running this business? Based on the information at www.candyusa.org answer the following questions:
1.) Which holiday will you want to have the most amount of candy available to sell? Why?
2.) During what months will your business be open longer hours? Why?
3.) If your store were to only sell chocolate, where in the world would you open your business? Why would you open your store in this country?

Expectation: Learners will work with a partner. Learners will answer these questions about their store based on the statistics available at www.candyusa.org Learners will write a journal entry detailing their answers.

BEFORE activity: Discuss why looking at statistics before making a decision in the business world can benefit a business.

Time allotment: Before ≈ 10 minutes **During** ≈ 10 minutes **After** ≈ 15 - 20 minutes.

Notes

possible hints:

key ideas for discussion:
How has the data changed over time?
conducting discussion:
before:

after:

specific groups or children from whom to gather data:

what to do with written work: examine for thoroughness of response (design a rubric for evaluation)

Linking to Key Teaching Strategies

Effective Mathematics Environment

Worthwhile Mathematical Task

Cooperative Learning

Models

Discourse, Writing and Active Listening
Letters to "stockholders" justifying choices could be substituted for journal entries and used as a vehicle for process writing skills.

CANDY STORE

Name: _____

You and a friend decide to open a candy store! Use the information on the www.candyusa.org website to answer the following questions:

Which holiday will you want to have the most amount of candy available to sell? Why?

_____
_____
_____

During what months will your business be open longer hours? Why?

_____
_____
_____

If your store were to only sell chocolate, where in the world would you open your business? Why would you open your store in this country?

_____
_____
_____

4. Garden Fence

Grade: 3-5 (can be modified to address what you know about your students).

Math: I want the learners to develop understanding of perimeter.

Task: Farmer Joe wants to put a fence around his square garden to keep the deer from eating the corn. One side of the fence will be 10m (meters) in length. If the posts need to be placed 2m apart, how many posts should he use?

Expectation: Have the students in pairs or groups of 3. Give students the problem to solve along with a geoboard and geobands, and/or dot paper (see Part II: Blackline Masters), or use a geoboard applet (see Van de Walle chapter 8 for some geoboard sites). Learners will explain their solutions to the problem in a journal entry. Discuss different solutions students found when they are all finished solving.

BEFORE activity: Generate a list of critical attributes of squares. Discuss reasons why people build fences in their yards.

Time allotment: Before ≈ 5 -10 minutes **During** ≈ 15 - 20 minutes **After** ≈ 15 - 20 minutes

---

Notes

possible hints:

key ideas for discussion:
Distance. For measuring the distance, what do we count— the "posts" or the "space" between the "posts?"
conducting discussion:
before: It is important to help the learners to recognize that if one side of the fence of a square garden is X units, then all four sides are X units each.

after:

specific groups or children from whom to gather data:

what to do with written work:

Linking to Key Teaching Strategies

Effective Mathematics Environment

Worthwhile Mathematical Task

Cooperative Learning

Models
This task is facilitated by the use of geoboards which can be used to model the garden fence.

Discourse, Writing and Active Listening
The before activity helps learners to reflect on their understanding of squares and will allow for the address of any potential misconceptions.

GARDEN FENCE

Name: _____

Farmer Joe wants to put a fence around his square garden to keep the deer from eating the corn. One side of the fence will be 10m (meters) in length. If the posts need to be placed 2m apart, how many posts should he use?

Use this space to solve this problem. You can draw pictures or write number sentences to help you. Be sure to explain how you know that your solution works!

5. Fraction Pictures

Grade: 3 - 6 (can be modified to address what you know about your students).

Math: I want the learners to develop fraction operation sense.

Task: Write the fraction problem and answer suggested by each picture. The units are the centimeter and square centimeter.

Expectation: Learners will work independently. Learners should be able to write at least one symbolic representation for each of the illustrations

BEFORE activity: Present a fraction number sentence (e.g., 1/2 + 2/3) and discuss what this might look like with pictures.

Time allotment: Before ≈ 5 - 10 minutes **During** ≈ 10 - 15 minutes **After** ≈ 10 - 15 minutes

Notes

possible hints:

key ideas for discussion:
How does the picture help to show why the answer to a multiplication of fractions less than one results in a product less than either of the factors? What does multiplication as an operation really mean?

conducting discussion:
before:

after:.

specific groups or children from whom to gather data:

what to do with written work:

Linking to Key Teaching Strategies

Effective Mathematics Environment

Worthwhile Mathematical Task

Cooperative Learning

Models
In this task, learners are taking a pictorial model of a mathematics sentence and attempting to translate it into a symbolic model. This type of activity can increase the level of what Van de Walle refers to as "relational understanding" (or more connections of "red" and "blue" dots).

Discourse, Writing and Active Listening

Fraction Pictures

Name: _____

Write the fraction problem, and answer, suggested by each picture. The units are the centimeter and square centimeter

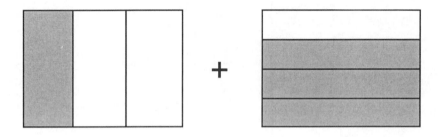

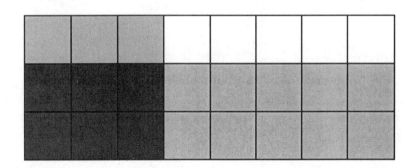

6. Factor Quest

Grade: 3-5 (can be modified to address what you know about your students).

Math: I want the learners to develop an understanding of the relationship between a number and its factors.

Task: How many different rectangular arrangements of X counters can you make?

Expectation: Using several numbers that have a relatively large number of factors, learners will attempt to build rectangular arrays that have the given number. Learners will write both addition sentences and multiplication sentences for each rectangle they make.

BEFORE activity: Pose the problem using the number 12 and have the learners find the different arrangements of counters and write the sentences.

Time allotment: Before ≈ 5 - 10 minutes **During** ≈ 15 - 20 minutes **After** ≈ 15 - 20 minutes

---

Notes

possible hints:

key ideas for discussion:

conducting discussion:
before:

after: How many arrays did you get for 36? Did you count the 6 x 6 array? Is there a relationship between the number and the number of arrays, and whether one of the arrays is a square?

specific groups or children from whom to gather data:

what to do with written work:

Linking to Key Teaching Strategies

Effective Mathematics Environment
As an activity that encourages investigation, this helps to support the environment characterized by "risk-free" problem-solving.

Worthwhile Mathematical Task

Cooperative Learning

Models
This activity draws connections between the concept factors with the rows and columns of arrays that can be made (and rearranged) from the specified number of counters.

Discourse, Writing and Active Listening

FACTOR QUEST

Name:_____

Use the grid here to record your arrangements. Use crayons to color them in. Be sure to label each arrangement with a multiplication sentence. (You might even write a division sentence!)

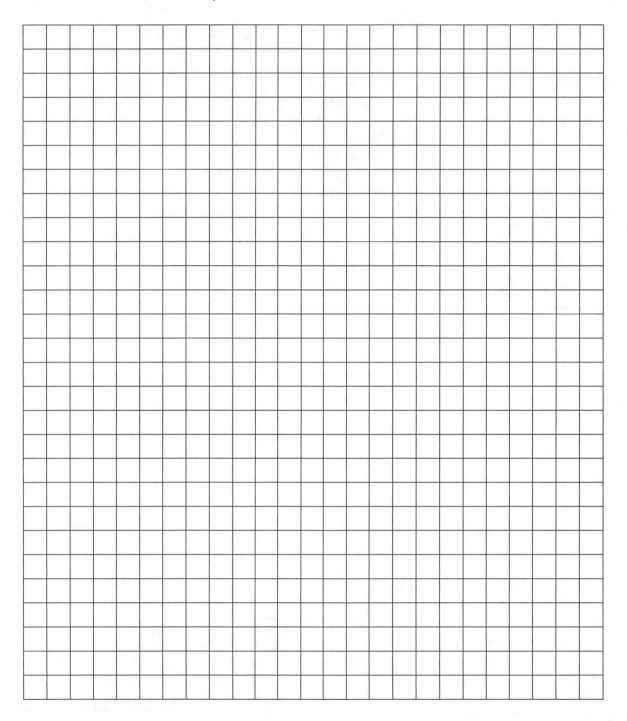

7. The Find!

Grade: PreK-2 (can be modified to address what you know about your students).

Math: I want the learners to develop base-ten numeration: grouping by ten and the role of positional digits.

Task: "I was looking in the closet and found this box of (some countable objects like beans, connecting cubes, etc.). How many do you think there are? Can you help me count them?"

Expectation: Learners will be given a set of counters of a random amount between 50 and 100. In groups of 3 - 4, learners will be asked to show how many there are and how they know. Learners will be asked to explain how to communicate that number to others (orally, written, symbolically).

BEFORE activity: Present the problem and brainstorm different strategies for counting them. Discuss which would be the most effective strategy.

Time allotment: Before ≈ 2-5 minutes **During** ≈ 5-10 minutes **After** ≈ 15 - 20 minutes.

Notes

possible hints:
You may need to facilitate movement from grouping by 2 or 5 to grouping by 10.

key ideas for discussion:
After they have determined how many, ask how someone else can know without having to count them over.

conducting discussion:
before:

after: Discuss the meaning of digits. What is the difference between writing 12 and 21? Relate this difference with spelling difference between *on* and *no* Have the children in each group write the number of objects they have as a numeral on 3x5 cards. Discuss the meaning of their numeral

specific groups or children from whom to gather data:

what to do with written work:

Linking to Key Teaching Strategies

Effective Mathematics Environment

Worthwhile Mathematical Task

Cooperative Learning

Models

Discourse, Writing and Listening Actively
Many opportunities to engage learners in discourse exist in this activity. When they are writing numerals, consider such questions as, "which is the first digit? the second digit? When does the first digit get changed? What would happen if we had 1 (2, 3, etc.) more groups of ten? When does the second digit get changed? What would happen if we had 1 (2, 3, etc.) more objects?

The Find!

Name: _____

Use this space to show how many counters
you have in your closet. You can use
pictures, words, or numerals.

8. Building Bridges
(Idea adapted from MSEB Assessment prototype)

Grade: 3 - 5 (can be modified to address what you know about your students).

Math: I want the learners to construct the relationship between two variables.

Task: We are bridge builders trying to determine the amount of material we need to construct our bridges. If we define a bridge span as (illustration 1), then how many uprights and how many cross pieces do we need to build an X(any number) span bridge?

Expectation: Working individually, learners will construct bridges of different span lengths for each type of bridge presented. Learners will determine a relationship to figure out how many pieces they need to build any bridge length. Learners will record their solutions on a recording sheet.

BEFORE activity: Present a simple bridge sample

Time allotment: Before ≈ 5 - 10 minutes **During** ≈ 20 minutes **After** ≈ 20 minutes.

Notes

possible hints:

key ideas for discussion:
Discuss how the two different types of bridges are similar and different

conducting discussion:
before:

after: Get the learners to talk about how they came up with their rules. How do they know their rules work?

specific groups or children from whom to gather data:

what to do with written work:

Linking to Key Teaching Strategies

Effective Mathematics Environment
This task is very much a problem-solving activity and can be used to promote a problem-solving environment. The availability of materials to suport this activity is part of an effective environment.

Worthwhile Mathematical Task

Cooperative Learning

Models
Although this task calls for the use of Cuisenaire® rods, it could be accomplished using any material, pipe cleaners, Unifix® cubes, blocks, etc. by ensuring that the uprights are the same size, and the cross pieces (for each bridge type) are the same length.

Discourse, Writing and Active Listening

BUILDING BRIDGES   Name:_____

Using Cuisenaire® rods, construct a one span bridge like this one:

Use the same color for the upright pieces.
Use any color for the cross piece

Now, using this type of span, build a 3 span bridge.
1) How many uprights did you need?
2) How many cross pieces did you need?
3) How long (in centimeters) is your bridge?
4) How many pieces did you need altogether?

Build a 5 span bridge.
1) How many uprights did you need?
2) How many cross pieces did you need?
3) How long (in centimeters) is your bridge?
4) How many pieces did you need altogether?

Without building a 9 span bridge,
1) How many uprights would you need? How do you know?
2) How many cross pieces would you need? How do you know?
3) How long (in centimeters) will your bridge be? How do you know?
4) How many pieces would you need altogether?

Write a rule for figuring out the total number of rods you would need to build a bridge if you knew how many spans the bridge had

_____

_____

Repeat the above task using the following type of bridge. This is a 2-span bridge.

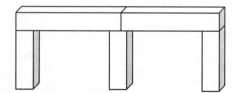

9. Ranking Theme Parks

Grade: 4 - 6 (can be modified to address what you know about your students).

Math: I want the learners to explore sampling data and compare statistics of different sample sizes.

Task: What are the most popular theme parks in North America? Survey 100 people to find out which theme park they would most like to visit. Compare your findings with those at:
http://enquirer.com/editions/2000/01/16/loc_top_50_north.html

Expectation: Working in teams, learners will devise a plan to collect the data on 100 people. As teams, they will report their findings and the comparison to the internet data in an oral presentation with write up to the class. They will explain why they think the results are similar or different.

BEFORE activity: Talk about surveys and what they are for.

Time allotment: Before ≈ 5 -10 minutes **During** ≈ may be days **After** ≈ 30 - 60 minutes.

Notes

possible hints:
You may need to help them consider how they will get data on 100 people who will be different from any other group.

key ideas for discussion:

conducting discussion:
before:

after: Compare the national survey's results with the class. Brainstorm about why they might be the same or different.

specific groups or children from whom to gather data:

what to do with written work:

Linking to Key Teaching Strategies

Mathematics Environment

Worthwhile Mathematical Task

This task appropriately represent the concepts and procedures related to survey statistics. It also involves the learners in "doing" mathematics by requiring them to design, administer, and interpret results of a survey. In doing so, it facilitates the development of these important skills.

Cooperative Learning

Models

Discourse, Writing and Active Listening

Ranking Theme Parks

What are the most popular theme parks in North America? Work with your team and make a plan to survey 100 people to find out which theme park they would most like to visit.

Collect your data and compare your findings with those at
http:// enquirer.com/editions/2000/01/16/loc_top_50_north.html

It may help to present your data here in a chart or graph. Be prepared to present your findings to the class.

Explain why you think your results are the same or different from those presented on the web site.

10. Stacking Blocks

Grade: 2 - 4 (can be modified to address what you know about your students).

Math: I want the learners to identify the pattern of triangular numbers.

Task: Lois and Tim are stacking blocks like little "stairways" (see picture). If they continue in the same pattern, how many blocks will they need to make a "stairway" nine steps high?

Expectation: Learners may work independently or with partners. Learners may draw or stack blocks to reason through the problem. Learners will explain how they know their solution is correct. Each learner will need at least 50 cubes.

BEFORE activity: Present the problem of "how many blocks to make a 'stairway' four steps high? Explain."

Time allotment: Before ≈ 10 minutes **During** ≈ 15 - 20 minutes **After** ≈ 15 - 20 minutes.

Notes

possible hints:

key ideas for discussion:

conducting discussion:
before:

after: Did it matter whether blocks were viewed right to left or up and down?

specific groups or children from whom to gather data:

what to do with written work:

Linking to Key Teaching Strategies

Effective Mathematics Environment
This activity relies on the availability of a variety of legitimate ways to approach the problem. While not necessarily a "real" situational problem, it has potential to pique learners curiosity and is based on sound and significant mathematics.

Worthwhile Mathematical Task

Cooperative Learning

Models

Discourse, Writing and Active Listening

Stacking Blocks

Name:_____

Lois and Tim are stacking blocks like little "stairways".

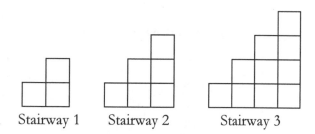

Stairway 1   Stairway 2   Stairway 3

If they continue in the same pattern, how many blocks will they need to make a "stairway" nine steps high?

Use the space below to draw pictures or write number sentences to help you solve this problem.

11. Cover All

Grade: 3 - 5 (can be modified to address what you know about your students).

Math: I want the learners to develop spatial sense, area concepts and practice estimation skills.

Task: How many square tiles (1 inch by 1 inch) are needed to cover this page?

Expectation: Learners may work in pairs to discuss the problem but must submit an individual write up. Learners may verify their estimates using real tiles. Learners will explain their solution process.

BEFORE activity: Conduct some estimation activities in one dimension involving non-standard units of length.

Time allotment: Before ≈ 5 - 10 minutes **During** ≈ 10 - 15 minutes **After** ≈ 10 - 15 minutes.

Notes

possible hints: Building on the **BEFORE activity** of length estimates, ask how many of the tiles wide or long the page is. What does this tell them with regard to rows and columns of an array model of multiplication?

key ideas for discussion:
the array model of multiplication and its relationship to area

conducting discussion:
before:

after:

specific groups or children from whom to gather data:

what to do with written work:

Linking to Key Teaching Strategies

Effective Mathematics Environment

Worthwhile Mathematical Task
This activity develops relevant skills of estimation and measurement. Can be made situational in the context of art classes doing mosaics.

Cooperative Learning

Models
This task may depend on the availability (or creation of) 1 in² tiles.

Discourse, Writing and Active Listening

Cover All

Name:_____

For a mosaic project in art, you and your friend are trying to determine how many tiles of this size will cover this page. How do you know? Explain your answer in the space below.

12. Interference

Grade: 4 - 8 (can be modified to address what you know about your students).

Math: I want the learners to develop least common multiple concepts.

Task: Two artificial satellites are in orbits that pass directly over your school. When they are both directly over your school, they cause interference with your schools telecommunication reception. One satellite makes one revolution around the Earth every 20 hrs; the other makes one revolution around the Earth every 30 hrs. At 8:00 A. M. on December 3 they were both directly over your school. When (date and time) will both next be directly over your school?

Expectation: Learners will work in pairs to discuss the problem but must submit an individual write up. Learners will explain their solution process. Learners may draw or use models to help them in their solution process.

BEFORE activity: Present the task and brainstorm as a class some approaches to the problem.

Time allotment: Before ≈ 5 - 10 minutes **During** ≈ 15 - 20 minutes **After** ≈ 20 minutes.

Notes

possible hints: Encourage the construction of a table to examine the relationships. Help them to be wary of the "passage of time" (i.e., 24 hrs to a day) -help them devise ways to keep track.

key ideas for discussion:
least common multiples

conducting discussion:
before: You might need to consider discussing the meaning of orbit.

after: Have learners explore their tables to look for patterns. Stimulate discussion around the idea that the time elapsed from the start time to each time a satellite is over the school is a multiple of the number of hours of the orbit.

specific groups or children from whom to gather data:

what to do with written work:

Linking to Key Teaching Strategies

Mathematics Environment

Worthwhile Mathematical Task
This activity focuses on a relevant phenomenon that has potential to pique learners interest.

Cooperative Learning

Models

Discourse, Writing and Active Listening
Getting learners to clearly define the problem and to discuss the notion of least common multiples depends on the environment for discourse in the classroom. Write ups could include a letter to the school board explaining when they can anticipate telecommunication interference.

Interference

Two artificial satellites are in orbits that pass directly over your school. When they are both directly over your school, they cause interference with your schools telecommunication reception. One satellite makes one revolution around the Earth every 20 hrs; the other makes one revolution around the Earth every 30 hrs. At 8:00 A. M. on December 3 they were both directly over your school. When (date and time) will both next be directly over your school?

Explain your answer in the space below. (Or write a letter to the school board indicating when they can expect the next telecommunication interference and how you know).

13. Triangle Communication

Grade: 3 - 5 (can be modified to address what you know about your students).

Math: I want the learners to develop language associated with the attributes of triangles.

Task: Your friend calls you on the phone asking for help. She needs to make an isosceles right triangle on her geoboard. What can you tell her over the phone to help her make an isosceles right triangle on her geoboard? Write your instructions below.

Expectation: Learners will work in pairs but submit individual responses. Learners will write clear instructions using appropriate vocabulary. Learners will use geoboards or dot paper (see black line masters) to help them in their solution process.

BEFORE activity: Have learners work in pairs. Have them place a partition of some sort (a large notebook, etc.) between them blocking the view of each other's geoboard (or dot paper). Learner 1 will make a triangle and describe to learner 2 how to make it. Learners can compare their work and trade roles to develop their communication skills and language.

Time allotment: Before ≈ 10 - 15 minutes **During** ≈ 15 - 20 minutes **After** ≈ 20 minutes.

Notes

possible hints:. For the isosceles right triangle, you may need to develop the vocabulary isosceles and right triangle.

key ideas for discussion:

conducting discussion:
before: Think about spatial orientation. How important is it to making a compatible triangle?

after:

specific groups or children from whom to gather data:

what to do with written work:

Linking to Key Teaching Strategies

Effective Mathematics Environment
Communication activities like this can foster the development of mathematical conjecture and argument. An environment that stimulates discussion can result in questions asked that reflect understanding and desire to learn more.

Worthwhile Mathematical Task

Cooperative Learning

Models

Discourse, Writing and Active Listening
Learners have the opportunity afterward to defend their descriptions of how to make an isosceles right triangle. (Note that this task can be modified for any particular shape constructed on a geoboard. The point is to develop the language necessary to convey the concept).

Triangle Communication

Name:_____

Your friend has just phoned you with a problem that has her "stumped." She needs to make an *isosceles right triangle* on a geoboard. Can you help her by giving her explicit instructions for how to do that?

Practice making one on your geoboard (or you can use dot paper) so that you are comfortable with how to do it. Then write out your instructions below. Share these instructions with a partner to see whether they are complete and accurate.

14. Tricking the Trickster

Grade: 2 - 8 (can be modified to address what you know about your students).

Math: I want the learners to express relationships symbolically or pictorially.

Task: Your older brother attempts to "trick" you with an old number trick. He says to think of a number, double it and add nine. Then add your original number to it and divide by three. Now add four and subtract your original number. The result, he says, is seven. How does this work? Can you write or draw the relationship between the result and the steps taken to get the result? Show your brother that this "trick" is really just "mathematics in action."

Expectation: Learners will work in pairs but submit individual responses. Learners will write clear descriptions of what happens at each step of the "trick." Learners will use symbols or pictures to help them in this process.

BEFORE activity: Present the trick to the learners initially as an opportunity to do some mental mathematics. Ensure that they all come up with the expected answer. Ask them whether they think it will work all the time. How do they know? Hand out the attachment and encourage them to work out an explanation for the "trick."

Time allotment: Before ≈ 10 - 15 minutes **During** ≈ 15 - 20 minutes **After** ≈ 20 minutes.

Notes

possible hints:.

key ideas for discussion:

conducting discussion:
before:

after: Ask how they know it works. Challenge them to make up a number "trick" on their own. Discuss algebraic representation connections.

specific groups or children from whom to gather data:

what to do with written work:

Linking to Key Teaching Strategies

Effective Mathematics Environment

Worthwhile Mathematical Task
Algebra is a formal symbolic language for mathematics. Learners in this task have an opportunity to discover the power of algebra as a communication tool. Being able to express the relationship either in symbols or in pictures lays a nice foundation for future algebraic concepts.

Cooperative Learning

Models

Discourse, Writing and Active Listening

Tricking the Trickster   Name:_____

Your older brother attempts to "trick" you with an old number trick. Here is the trick:

1. Choose a number
2. Double it.
3. Add nine
4. Add your original number
5. Divide by three
6. Add four
7. Subtract your original number
8. Your result is seven.

Did it work? Does it work every time? How do you know? Can you write or draw the relationship between the result and the steps taken to get the result? Show your brother that this "trick" is really just "mathematics in action."

15. Odd or Even?

Grade: 1 - 3 (can be modified to address what you know about your students).

Math: I want the learners to develop number sense and concepts of odd and even.

Task: Is the sum of two consecutive numbers odd or even? How do you know? Write an explanation.

Expectation: Learners may work in pairs to discuss the problem but must submit an individual write up. Learners may use counters to help them. Learners will complete a table of samples sums. Learners will explain their solution process.

BEFORE activity: Explore characteristics of odd and even numbers. Even numbers can form pairs with no "left overs." Odd numbers can form pairs with one "left over" or "an odd person out."

Time allotment: Before ≈ 5 - 10 minutes **During** ≈ 10 - 15 minutes **After** ≈ 10 - 15 minutes.

Notes

possible hints: You may need to help these learners see that for any two consecutive numbers, one is even and one is odd.

key ideas for discussion:
odd and even concept

conducting discussion:
before:

after: Ask the learners to share the results of their tables and to share any patterns they discovered.

specific groups or children from whom to gather data:
Which children are able to describe that even plus odd makes pairs, and pairs with one left over which is odd?

what to do with written work:

<table>
<tr><td colspan="2"><u>Linking to Key Teaching Strategies</u></td></tr>
<tr><td>Effective Mathematics Environment</td></tr>
<tr><td>Worthwhile Mathematical Task</td></tr>
<tr><td>Cooperative Learning</td></tr>
<tr><td>Models
This activity provides an opportunity to use counters to model the "pairing" concept of even numbers (as well as the "pairing with one left out" concept of odd numbers). Through this kind of modeling, learners may see the concept of consecutive number sums and may even be able to represent it symbolically or pictorially, or be able to describe in words.</td></tr>
<tr><td>Discourse, Writing and Active Listening</td></tr>
</table>

Odd or Even?

Name:_____

Is the sum of two consecutive numbers odd or even? How do you know? Complete the table below.

first addend	second addend	sum

Are the sums odd or even? Do you think the sums will always be that way? Write an explanation for how you know.

16. Hiking

Grade: 4 - 6 (can be modified to address what you know about your students).

Math: I want the learners to estimate and indirectly measure length.

Task: Cicely and her friends went hiking. They started from Jasper Point and hiked the trail to Friends Pond. How far did they hike?

Expectation: Learners will use measuring tools appropriately to measure the length of the path on the map and convert the distance from the scale to actual distance. Learners will work together in pairs to solve the problem. Learners will present their solutions to the class.

BEFORE activity: Learners will be asked to measure "atypical" lines and curves such as

Time allotment: Before ≈ 10 minutes **During** ≈ 20 minutes **After** ≈ 20 minutes.

Notes

possible hints:
Is there anything that could be "bent" around these paths and then straightened to measure the distance?

key ideas for discussion:
Distances to be measured are not always "straight" lines.

conducting discussion:
before:

after: Can you explain why your method of measurement works?

specific groups or children from whom to gather data:

what to do with written work:

Linking to Key Teaching Strategies

Effective Mathematics Environment

Worthwhile Mathematical Task
This particular task attends to potential "real life" situations and has potential to pique learner interest. It can be approached from a variety of interesting and legitimate ways and fosters the relevant skill development of measurement.

Cooperative Learning

Models

Discourse, Writing and Active Listening

Hiking

Name:_____

Cicely and her friends went hiking. They started from Jasper Point and hiked the trail to Friends Pond. How far did they hike? Explain, in writing, your solution process.

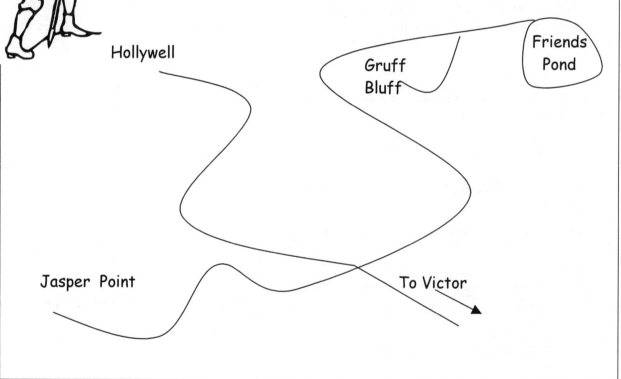

Scale: 1 inch = half mile

17. Cakes, Cakes, and More Cakes

Grade: 1 - 2 (can be modified to address what you know about your students).

Math: I want the learners to add with multiple addends

Task: Hannah is bringing cupcakes to school tomorrow for an after school class party. Her dad is baking them and Hannah is going to put them in boxes. There are 36 cupcakes on the counter, 17 on a plate (there was 18 but her brother ate one), and 18 more in the oven. How many cupcakes will Hannah have to box?

Expectation: Learners can use manipulative models to represent the cupcakes. Learners will describe how they approached the problem.

BEFORE activity: Brainstorm ways to approach this problem.

Time allotment: Before ≈ 5 - 10 minutes **During** ≈ 15 - 20 minutes **After** ≈ 15 - 20 minutes

---

Notes

possible hints:
Try grouping models into sets of ten.

key ideas for discussion:

conducting discussion:
before:

after:

specific groups or children from whom to gather data:

what to do with written work:

Linking to Key Teaching Strategies

Effective Mathematics Environment

Worthwhile Mathematical Task

Cooperative Learning

Models

Discourse, Writing and Listening Actively
This activity promotes the thinking and reasoning associated with multiple addend addition. This thinking can be revealed by the descriptions learners give for their approaches. You can observe their use of models to determine whether they are systematically combining numbers or are "lumping" all together and counting one by one.

Cakes, Cakes, and More Cakes

Name:_____

Hannah is bringing cupcakes to school tomorrow for an after school class party. Her dad is baking them and Hannah is going to put them in boxes. There are 36 cupcakes on the counter, 17 on a plate (there was 18 but her brother ate one), and 18 more in the oven. How many cupcakes will Hannah have to box?

Use base - ten manipulatives to help you model this problem. Use the space below to describe how you solved this.

18. Life with Fractions

Grade: 4 - 6 (can be modified to address what you know about your students.

Math: I want the learners to develop number sense with fractions.

Task: Liam and his friend Thomas are both trying to decide whether $2\frac{1}{3} \times \frac{3}{8}$ should be more or less than 1. Thomas thinks it should be more than one. Liam thinks it should be less than one. Who do you agree with? Why? Explain.

Expectation: Learners will explore this idea before being exposed to any fraction computational "rules." They will share their reasoning with the class. They might choose to use manipulative models.

BEFORE activity: Review the meaning of multiplication as combining sets of equal size. For example, 2×3 means 2 "sets" (or groups) of 3 things.

Time allotment: Before $\approx$ 5 minutes **During** $\approx$ 5 - 10 minutes **After** $\approx$ 10 - 15 minutes.

Notes

possible hints:

key ideas for discussion:

conducting discussion:
before:

after: Elicit the learner's conjectures and justification for why they think Liam or Thomas is correct. Ask how they used (or might use) a model.

specific groups or children from whom to gather data:

what to do with written work:

<table>
<tr><td colspan="2"><u>Linking to Key Teaching Strategies</u></td></tr>
<tr><td>Effective Mathematics Environment
This question focuses attention on the development of number sense rather than computational algorithms, and consequently promotes reflective thinking among the learners and making conjectures- key elements of an effective environment.</td></tr>
<tr><td>Worthwhile Mathematical Task</td></tr>
<tr><td>Cooperative Learning</td></tr>
<tr><td>Models</td></tr>
<tr><td>Discourse, Writing and Active Listening</td></tr>
</table>

Life with Fractions

Name:_____

Liam and his friend Thomas are both trying to decide whether
$2\frac{1}{3} \times \frac{3}{8}$ should be more or less than 1. Thomas thinks it should
be more than one. Liam thinks it should be less than one. Who
do you agree with? Why? Explain.

19. Squeeze!

Grade: 3- 5 (can be modified to address what you know about your students).

Math: I want the learners to develop fraction number sense.

Task: Find a fraction that is between $\dfrac{5}{8}$ and $\dfrac{3}{4}$.
Explain, or show, how you know you're right.

Expectation: Learners will work with a partner and write as many explanations as they can. They will share their explanations with the class and decide whether they make sense.

BEFORE activity: "Warm up" with some concrete models of fractions, comparing physical size and connecting to symbolic representation.

Time allotment: Before ≈ 5 - 10 minutes **During** ≈ 10 - 15 minutes **After** ≈ 10 - 15 minutes.

Notes

possible hints:

key ideas for discussion:

conducting discussion:
before:

after: Learners can be directed to write their explanations on the overhead or the board (or you can do so) and each explanation can be discussed as to whether it makes sense.

specific groups or children from whom to gather data:

what to do with written work:

Linking to Key Teaching Strategies

Effective Mathematics Environment
This kind of task presents the learners with the opportunity to explore meaning in mathematics without a dependence on "rules." This can foster an atmosphere of thinking and reasoning in mathematics.

Worthwhile Mathematical Task

Cooperative Learning

Models

Discourse, Writing and Active Listening

Squeeze!

Name:_____

Find a fraction that is between $\frac{5}{8}$ **and** $\frac{3}{4}$. **How do you know you are right?**

How many different ways of explaining this can you find?
Record all the ways that you and your partner find below.

20. Illustrating Ratios

Grade: 3 - 6 (can be modified to address what you know about your students).

Math: I want the learners to graph equal ratios to predict other equal ratios.

Task: Monique works for a company that makes Doohickeys. Her company has one machine that can make a certain number of Doohickeys per minute. Monique observed the machine and collected data to see how many Doohickeys were made after each minute. An order for 72 Doohickeys came in from AB textbook publishers. She needs to predict how long it will take to make the 72 Doohickeys. From the data provided, can you help her make her prediction? How long will it take to make 100 Doohickeys?

Expectation: Learners will complete the chart and graph the data to make a prediction. They will write a letter to the AB textbook publisher explaining how long it will take to complete their order and how they know it will take that long.

BEFORE activity: Practice completing some other scaling activities. (See Van de Walle, Chapter 18)

Time allotment: Before ≈ 10 minutes **During** ≈ 15 - 20 minutes **After** ≈ 15 - 20 minutes

---

Notes

possible hints:

key ideas for discussion:

conducting discussion:
before:

after: Have learners present their graphs and explain why they did as they did.

specific groups or children from whom to gather data:

what to do with written work: Letters can be posted on a bulletin board or placed in learner portfolios as evidence of understanding.

Linking to Key Teaching Strategies

Mathematics Environment

Worthwhile Mathematical Task

Cooperative Learning

Models

Discourse, Writing and Active Listening
This activity offers the opportunity to incorporate connections to writing. By having the learners write a letter that explains the problem solution, they not only present evidence of their understanding of the mathematics, they present evidence of understanding the processes involved in writing letters (literacy connection).

Illustrating Ratios

Name:_____

Monique works for a company that makes Doohickeys. Her company has one machine that can make a certain number of Doohickeys per minute. Monique observed the machine and collected data to see how many Doohickeys were made after each minute. An order for 72 Doohickeys came in from AB textbook publishers. She needs to predict how long it will take to make the 72 Doohickeys. From the data provided, can you help her make her prediction? How long would it take to make 100 Doohickeys?

Minutes	4			16	22	28	34
Doohickeys			27	36			

Graph your results here. Label the axes.

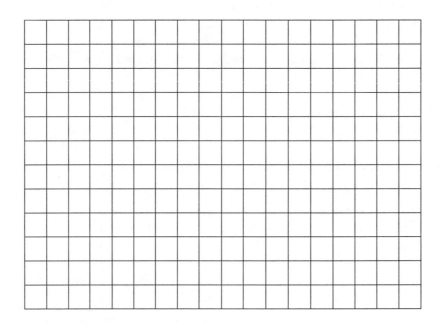

How long will it take to make 72 Doohickeys? How do you know?

How long will it take to make 100 Doohickeys? How do you know?

Write a letter to AB textbook publishers explaining how long it will take to make their 72 Doohickeys and how you know. Use proper letter writing style.

21. Right Angle Challenge

Grade: 5 - 8 (can be modified to address what you know about your students).

Math: I want the learners to develop conceptual understanding of Pythagorean Triples.

Task:. Ancient Egyptians would create right angles from a length of rope to help them in reestablishing land boundary lines after periodic flooding of the Nile. Can you and two partners figure out how to make a right angle with just a length of rope?

Expectation: Learners will work in groups of three with a 12 foot length of rope. They will explore ways to make a right angle. They should be able to explain how they know that they created a right angle. They will share their solutions with the class.

BEFORE activity: Explore/review the relationship among the sides of a right triangle (Pythagorean Theorem).

Time allotment: Before ≈ 5 - 10 minutes **During** ≈ 10 - 15 minutes **After** ≈ 10 - 15 minutes

---

Notes

possible hints:

key ideas for discussion:
Measurement of the rope (i.e., it is 12 feet long, but could different Pythagorean Triples be made from it? Not just 3-4-5?)

conducting discussion:
before.

after. Are there other triples we could make with this rope?

specific groups or children from whom to gather data:

what to do with written work:

Linking to Key Teaching Strategies

Effective Mathematics Environment

Worthwhile Mathematical Task
This task presents an opportunity for a "practical" application of the Pythagorean Theorem. The context of the problem and the "challenge" may pique the learners interest and provide motivation to find a solution.

Cooperative Learning

Models
The use of the rope presents an historical context for this problem.

Discourse, Writing and Active Listening

Right Angle Challenge

Name:_____

Ancient Egyptians would create right angles from a length of rope to help them in reestablishing land boundary lines after periodic flooding of the Nile. Can you and two partners figure out how to make a right angle with just a length of rope?

Describe your solution below.

22. Rename That Number!

Grade: 2 - 4 (can be modified to address what you know about your students).

Math: I want the learners to develop ability to "rename" numbers from standard representations into other representations.

Task:. Carlo is trying to figure out how to represent 87, 195, and 212 using base-ten models. His set of base-ten models, however, only has 1 hundred square (flat), 7 tens (longs or rods) and 42 singles. How can he model these numbers with this set of pieces? Can you help him do this? How many different ways could you represent these numbers?

Expectation: Learners will work independently to find equivalent representations. Learners will record pictorial representations of the concrete equivalent representation.

BEFORE activity: Start as a class with 87. Brainstorm ideas for how to represent this number.

Time allotment: Before ≈ 5 - 10 minutes **During** ≈ 15 - 20 minutes **After** ≈ 15 - 20 minutes

---

Notes

possible hints:

What makes a hundred? a ten?

key ideas for discussion:

conducting discussion:
before:

after. Were there any of these numbers that could not be represented in more than one way with this set? Why?

specific groups or children from whom to gather data:

what to do with written work:

Linking to Key Teaching Strategies

Effective Mathematics Environment

Worthwhile Mathematical Task

Cooperative Learning

Models
These materials support the idea that mathematics is not in the material but in the relationships we form. This becomes evident in seeing that learners *can* show multiple representations of a number and not be "stuck" on the standard representation.

Discourse, Writing and Active Listening

Rename That Number!   Name:_____

Carlo is trying to figure out how to represent these numbers using base-ten models. 87, 195, and 212. His set of base-ten models, however, only has 1 hundred square (flat), 7 tens (longs or rods) and 42 singles. How can he model these numbers with this set of pieces? Can you help him do this? How many different ways could you represent these numbers?

Record your representations below using these base – ten model symbols:

☐	Carlo's set of
									base – ten models
• •	

23. Fruit Punch

Grade: 3 - 6 (can be modified to address what you know about your students).

Math: I want the learners to graph equal ratios to predict other equal ratios.

Task:. Kelley needs to know how many cans of fruit punch concentrate it will take to make 72 quarts of fruit punch for a class picnic. From the data provided, can you help her figure out her problem? Suppose she needed to make 78 quarts. How much concentrate would she need?

Expectation: Learners will complete the chart and graph the data to make a prediction. They will write a letter to Kelley explaining how much concentrate is needed and how they know.

BEFORE activity: Practice completing some other scaling activities (see Van de Walle, chapter 18 for ideas).

Time allotment: Before ≈ 5 - 10 minutes **During** ≈ 10 - 15 minutes **After** ≈ 10 - 15 minutes.

Notes

possible hints:

key ideas for discussion:

conducting discussion:
before:

after: Have learners present their graphs and explain why they did as they did.

specific groups or children from whom to gather data:

what to do with written work: Letters can be posted on a bulletin board or placed in learner portfolios as evidence of understanding.

Linking to Key Teaching Strategies

Mathematics Environment

Worthwhile Mathematical Task

Cooperative Learning

Models

Discourse, Writing and Active Listening
This activity offers the opportunity to incorporate connections to writing. By having the learners write a letter that explains the problem solution, they not only present evidence of their understanding of the mathematics, they present evidence of understanding of the processes involved in writing letters (literacy connection).

Fruit Punch

Name:_____

Kelley needs to know how many cans of fruit punch concentrate it will take to make 72 quarts of fruit punch for a class picnic. From the data provided, can you help her figure out her problem? Suppose she needed to make 78 quarts. How much concentrate would she need?

Cans of Concentrate	2			11	14	17	20
Quarts of Fruit Punch			32	44			

Graph your results here. Label the axes.

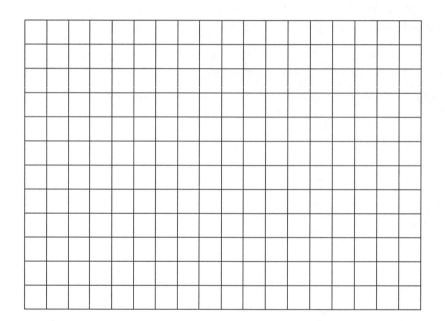

How many cans of concentrate are needed to make 72 quarts of fruit punch? How do you know?

How many cans of concentrate are needed to make 78 quarts of fruit puch? How do you know?

Write a letter to Kelley explaining your results and how you know. Use proper letter writing style.

24. Tangram Polygons

Grade: 3 - 6 (can be modified to address what you know about your students).

Math: I want the learners to develop spatial sense and explore relationships among polygons.

Task:: (non-situational) Using tangrams (either from the black line masters or from the tangram creation activity on p. 121), complete the chart below. Can you find a way to make each of the polygons listed below using 1 tangram piece, then 2 tangram pieces, then 3 pieces, and so on up to all seven pieces? How many different ways can you find for each?

Expectation: Learners will complete the chart by drawing their solutions. Learners may work in pairs or small groups. Learners will share their solutions with the class. Learners should be able to discuss why they think any shape can't be made with a particular number of tangram pieces.

BEFORE activity: Do a few together to model the expectations

Time allotment: Before ≈ 5 - 10 minutes **During** ≈ 10 - 15 minutes **After** ≈ 10 - 15 minutes.

Notes

possible hints:

key ideas for discussion:

conducting discussion:
before:

after: Were there any shapes and number of tangram pieces for which you were unable to make? Do you think there is a reason for that? Why?

specific groups or children from whom to gather data:

what to do with written work:

Linking to Key Teaching Strategies

Effective Mathematics Environment
This activity focuses the learner on the development of spatial sense. It encourages conversation about strategies among learners as they attempt to create polygons from the tangram pieces. Trying to find multiple solutions can contribute to the development of a disposition for making conjectures. That is, learners who are aware that there may be more than one possible way and are encouraged to find them may be inclined to seek other possibilities in other situations.

Worthwhile Mathematical Task

Cooperative Learning

Models

Discourse, Writing and Listening Actively

Tangram Polygons

Name:_____

Using tangrams, complete the chart below.

Can you find a way to make each of the polygons listed below using 1 tangram piece, then 2 tangram pieces, then 3 pieces, and so on up to all seven pieces? How many different ways can you find for each? Draw pictures of your solutions in the chart

Can you use tangram pieces to solve this problem?

Shape	2 pieces	3 pieces	4 pieces	5 pieces	6 pieces	7 pieces
Triangle						
Trapezoid						
Square						

25. Target Number

Grade: 1 - 8 (can be modified to address what you know about your students)

Math: I want the learners to develop their number sense and operation sense. I want the learners to "re-name" numbers.

Task: (non-situational) Roll 7 number cubes (two the same color, the rest a different color). For the two number cubes of the same color, multiply the value showing on one by ten then add to the other number cube. This becomes your target number. The values showing on the remaining 5 number cubes are then to be combined using each once and any mathematical operation known to you (including using some as exponents) to come as close to the target number as possible.

Expectation: Learners can work in small groups and each try to see who can come the closest. Older learners, presumably having more math content knowledge and skill, should be able to reach the target number exactly in many cases. For each combination, learners should record the expression and check for accuracy in both computation and expression. Learners will justify their expressions.

BEFORE activity: Do one as an example (e.g., a roll of 6 and 4 on the two same color cubes, and 2, 2, 3, 5, and 1 on the remaining cubes. The target number is $6 \times 10 + 4 = 64$. Then the expression could be $(2 \times 2)^3 \times 1^5 = 64$).

Time allotment: Before $\approx$ 5 - 10 minutes **During** $\approx$ 15 - 20 minutes **After** $\approx$ 15 - 20 minutes.

Notes

possible hints:
If necessary, suggest that they find the units digit first.

key ideas for discussion:
Different ways to express numbers symbolically.

conducting discussion:
before:

after: Ask them to defend their expressions.

specific groups or children from whom to gather data:

what to do with written work:

Linking to Key Teaching Strategies

Effective Mathematics Environment

Worthwhile Mathematical Task
Each target number can be approached from a variety of interesting and mathematically legitimate ways. In fact, there may be several different expressions for the same target number.

Cooperative Learning

Models

Discourse, Writing and Listening Actively
The opportunity for learners to hear others give their expressions and to compare them to their own provides ample opportunity to deepen their number and operation senses.

Target Number

Name:_____

Work in groups of 3 or 4. Use 7 number cubes. Mark two of them (or ensure they are the same color). Roll those two. Multiply the value showing on one of them by 10 and add it to the value showing on the other. This becomes your target number. Record this number on the chart below. Roll the other 5 number cubes. Use the values showing, each only once, to write an expression that evaluates to a number as close to the target number as you can (you might even hit the target number!). Record your expression in the chart below. Decide who in your group got the closest.

Target Number	My Expression	Who got the closest?	Closest Expression

26. Up to 57

Grade: 2 - 5 (can be modified to address what you know about your students).

Math: I want the learners to use tables or charts to organize information to solve problems. I want the learners to be able to rename monetary values using different denominations.

Task: Using our monetary system's coins, how many different ways can you make 57 cents? Which combination uses the most coins? Which uses the least? Are there any that use the same number of coins? How do you know that you have accounted for all combinations?

Expectation: Learners will work in pairs. Learners should have access to coin models. Learners will complete a chart or table with all possible combinations. Learners will analyze the information and answer the questions posed.

BEFORE activity: Pose the question and brainstorm ways to approach it.

Time allotment: Before ≈ 5 - 10 minutes **During** ≈ 10 - 15 minutes **After** ≈ 10 - 15 minutes.

Notes

possible hints:
You may need to encourage them to use a table or a chart.

key ideas for discussion:

conducting discussion:
before.

after. Why do we have the coins we do? Why don't we just have pennies, dimes, and dollars?

specific groups or children from whom to gather data:

what to do with written work:

Linking to Key Teaching Strategies

Mathematics Environment
In the guise of making change, this task promotes a mathematics environment that encourages exploration and investigation by posing the question of how many different ways there are to make 57 cents. This puts the learner in the position of planning an organized approach to the problem and devising ways to keep track of which combinations have been tried and which work. This promotes a problem solving environment.

Worthwhile Mathematical Task

Cooperative Learning

Models

Discourse, Writing and Active Listening

Up to 57

Name:_____

Using our monetary system's coins, how many different ways can you make 57 cents? As you work this out, which combination uses the most coins? Which uses the least? Are there any that use the same number of coins? How do you know that you have accounted for all combinations?

Use the space below to carry out your plan and show the combinations

How many ways can you make 57 cents?

27. Mean Basketball!

Grade: 2 - 6 (can be modified to address what you know about your students)

Math: I want the learners to explore measures of central tendency.

Task: The heights of the members of a basketball team are 5 ft 10 in, 5 ft 11 in, 6 ft 3 in, 6 ft 4 in, 6 ft 6 in, 6 ft 6 in, 6 ft 10 in, 7 ft 0 in, 7 ft 1 in, 7 ft 4 in. Find the average height of the team by finding the mean, median and mode. Which measure best represents the "average" of team's heigh? Explain your answer and provide a justification.

Expectation: Learners will work in pairs. Learners should discuss which measure of central tendency is best suited to this data set and why.

BEFORE activity: Review and discuss the similarities and differences among the three measures of central tendency: mean, median and mode.

Time allotment: Before ≈ 5 - 10 minutes **During** ≈ 10 - 15 minutes **After** ≈ 10 - 15 minutes

---

Notes

possible hints:

key ideas for discussion:

conducting discussion:
before: What do we mean by "average?"

after: Why do you think that particular measure is the best?

specific groups or children from whom to gather data:

what to do with written work:

Linking to Key Teaching Strategies

Effective Mathematics Environment

Worthwhile Mathematical Task

Cooperative Learning

Models

Discourse, Writing and Active Listening
The after portion of this activity provides an opportunity for learners to make conjectures regarding this idea. As a teacher, you have the opportunity to assess their understanding through the types of ideas they raise as to why they either agree or disagree. This will provide valuable information for future instructional plans.

Mean Basketball!

Name:_____

The heights of the members of a basketball team are:

> 5 ft 10 in, 5 ft 11 in, 6 ft 3 in, 6 ft 4 in, 6 ft 6 in,
> 6 ft 6 in, 6 ft 10 in, 7 ft 0 in, 7 ft 1 in, 7 ft 4 in.

Find the average height of the team by finding the mean, median and mode.

Mean Height   _____

Median Height   _____

Modal Height   _____

Which measure best represents the "average" height of the team?

_____.

Explain and justify your answer.

Assessment Tasks and Rubrics from Balanced Assessment for the Mathematics Curriculum

The following tasks and supporting materials were developed by the project Balanced Assessment for the Mathematics Curriculum and published in a series of **Assessment Packages** by Dale Seymour Publications, 2000. Further information about additional packages may be obtained from the publisher. Further information about additional tasks and supporting instructional and professional development materials may be obtained from the Balanced Assessment-Mathematics Assessment Resource Service project website: http://www.educ.msu.edu/mars/

These tasks and rubrics are offered as an example of how problem-based tasks can be used to look for learner understanding. The tasks themselves have been field tested with learners in public schools whose responses have been examined to develop the included rubrics. These rubrics are holistic in nature (see Van de Walle, page 68) and are intended to help the reader recognize the importance of identifying the core mathematics content and looking for evidence of *understanding* by describing elements of *performance*.

Each task also has a completed "sample" solution.

The first two tasks are designed for elementary level learners and can each be modified to address specific needs. The last task is designed for middle level learners and can also be modified.

You are encouraged to examine the descriptions of the rubrics and are invited to consider how performance in each of the preceding activities might be characterized.

For presentation in this guide, the original formatting of these tasks has been modified to fit the problem – based lesson format used by the preceding lessons and activities. On the lesson plan pages, the text that is not in bold characters are direct quotes from the Balanced Assessment task. In the side bar, notes that are taken directly from these tasks are quoted and cited. The accompanying activity sheets each contain the exact text and images from the original material, but have been reformatted to fit this guide. The sample solutions and descriptions of rubrics also contain the exact text from the original material but, they also have been reformatted to fit this guide.

28. Bolts and Nuts!

Source: Balanced Assessment for the Mathematics Curriculum*

Grade: 3 - 6 (can be modified to address what you know about your students).

Math: I want the learners to: Take pairs of measurements in a practical situation. Make simple ratio calculations.

Task: Students estimate, measure, and calculate the number of turns made to the nut and the distance it moves. Calculations bring in simple ideas of ratio, which pupils of this age will normally find quite challenging.

Expectation: Students should have met the idea of ratio in a practical context. They should be familiar with measurements in millimeters and meters... Students may discuss the task in pairs, but each student should complete an individual written response.

BEFORE activity: Arrange for students to work in pairs. Each student needs a calculator, a ruler, and a nut and bolt, which should not be connected together.

Demonstrate to the class how the nut can be attached to the bolt by twisting it on. Show how the nut can be moved along the bolt by turning it.

Ask students to try this. Encourage them to help each other. offer as much help as you can to anyone who finds this difficult. Give them a little time to play, as this is important in order to understand the task.

Now demonstrate how to place the ruler next to the bolt and count the number of turns needed to move the nut a distance of 10 millimeters. It is important that they understand that "complete turn" means one complete revolution of the nut. Demonstrate this to them making use of the marks on the nut and bolt.

Now hand out the activity pages and read them with the class to make sure they understand what they have to do. Explain that everyone must hand in an individual response.

Time allotment: Before ≈ 10 minutes **During** ≈ 35 minutes **After** ≈ X minutes.

Notes
materials: calculator, ruler nut and bolt
possible hints:
The nut and bolt should not be connected together. The bolt should be at least 5 cm long and no more that 9 cm long.

To help students count the turns, make a small mark with white correction fluid or a black permanent marker on one face of the nut and on one side of the head of the bolt.

key ideas for discussion:

conducting discussion:

before:

after:

specific groups or children from whom to gather data:

what to do with written work:

Linking to Key Teaching Strategies

Effective Mathematics Environment
As a problem task, this assessment activity is reflective of a problem- solving environment.

Worthwhile Mathematical Task
According to the criteria for worthwhile mathematical tasks (see part I), this task is worthwhile. It presents sound and significant mathematics in the context of an interesting problem that can be approached in a variety of interesting and legitimate ways.

Cooperative Learning
Although designed to assess individual reporting, the fact that it encourages initial discussion in pairs acknowledges the power of group learning.

Models
This task depends on the actual use of a concrete model (i.e., a real bolt and nut). Learners have the opportunity to actually experiment with the problem solution.

Discourse, Writing and Active Listening
The nature of the student response activity as well as the before activity, have potential to engage the learners in both meaningful discourse and writing.

The request in question four to "Explain how you figured this out" is intended to encourage learners to justify their answer. Depending upon the responses, you may have to think about how to prepare learners to understand what this means. Many learners (without such preparation) often respond with a procedural description rather than a conceptual one.

Bolts and Nuts!

Name: _____

Work with a partner on this problem.
You should have a nut, a bolt, a ruler, and a calculator.

Fix the nut onto the bolt.

Turn the nut so that it moves along the bolt.

1. How many complete turns will move the nut 25 mm along the bolt?
    First make a guess.
    Your guess: _____turns move the nut 25 mm.
    Your partner's guess: _____turns move the nut 25 mm.

    Talk to your partner about how you will answer this question. You will need to
    measure 25 mm with the ruler. There are some marks on the nut and bolt that
    may help you count the turns.

    Your answer: _____turns move the nut 25 mm.

    Your partner's answer:_____ turns move the nut 25 mm.

    Do you agree? If not, who has the right answer?

Now try to answer the following questions *without touching the nut and bolt.*

2.   How many times would you have to turn the nut to move it 50 mm?_____

3.   Complete this table.

Distance the Nut Moves	Number of Turns	
25 mm		<--- Write in your answer from the previous page.
50 mm		<--- Write in your answer from the last question.
100 mm		<--- Figure this out.
200 mm		<--- Figure this out.
1 meter		<---Figure this out.

To answer these you will need to imagine a bolt that is longer than the one you are using. The 1-meter bolt would be giant-sized because 1 meter = 1000 mm.

4.   How far would the nut move if you turned it 50 complete turns? Explain how you figured this out:

_____

_____

_____

Bolts and Nuts!   Sample Solution

1. It takes 20 turns to move the nut 25 mm.
2. You'd have to turn the nut about 40 times to move it 50 mm
3.

Distance the Nut Moves	Number of Turns
25 mm	About 20
50 mm	About 40
100 mm	About 80
200 mm	About 160
1 meter	About 800

4. If 1 meter or 1000 mm takes about 800 turns, then 125 mm takes about 100 turns. Therefore, 50 turns would move the nut about 62 mm.

Rubric

Characterizing Performance

This section offers a characterization of student responses and provides indications of the ways in which the students were successful or unsuccessful in engaging with and completing the task. The descriptions are keyed to the *Core Elements of Performance*. Our global descriptions of student work range from "The student needs significant instruction" to "The student's work meets the essential demands of the task."

The characterization of student responses for this task is based on these *Core Elements of Performance*:
1. Take pairs of measurements in a practical situation
2. Make simple ratio calculations

Descriptions of Student Work

The student needs significant instruction.
These papers show, at most, an attempt to make the estimates or measurements

The student needs some instruction.
An attempt has been made to make the estimates and measurements. The table has been at least partially completed, but there is a poor understanding of proportion.

The student's work needs to be revised.
Measurements have been correctly made within the generous margin of error (± 4 turns). The table has been partially completed with most figures in proportion.

The student's work meets the essential demands of the task.
Measurements have been correctly made within the generous margin of error (± 4 turns). The student may have indicated somewhere on the response that figures are "rough," "approximate," or "estimated." The student can handle simple ratio calculations as indicated either by the table completed with all figures in proportion, or by a consistent answer to the final question.

*This task comes from **Elementary Grades Assessment Package 2,** Dale Seymour Publications, 2000 (ISBN: 0-7690-0064-9). The task and supporting materials were developed by the project Balanced Assessment for the Mathematics Curriculum. Further information about additional packages may be obtained from the publisher. Further information about additional tasks and supporting instructional and professional development materials may be obtained from the Balanced Assessment-Mathematics Assessment Resource Service project website: http://www.educ.msu.edu/mars/*

29. Magic Age Rings

Source: Balanced Assessment for the Mathematics Curriculum*

Grade: 3 - 6 (can be modified to address what you know about your students).

Math: **I want the learners to:** Use simple mathematical functions. Explore and discuss the importance of order of operations and the relationship of addition, subtraction, multiplication, and division. Combine various arithmetic operations to solve complex problems.

Task: This task asks students to imagine wearing magic rings that change their age. A blue ring doubles one's age, a green ring adds five years, and a yellow ring takes two years away from one's age. Students calculate the effects of wearing different rings and answer questions about this imaginary situation.

Expectation: Most fourth graders should be able to tackle the mathematics of this task. It assumes that students have had prior opportunities to explore the ways addition, subtraction, multiplication, and division relate to each other. The task also assumes that students have some understanding of the significance of order of operations and that they know the meaning of "doubling" a quantity…Students may discuss the task in pairs, but each student should complete an individual written response.

BEFORE activity: Hand out the task prompt to students and read the first page aloud to them. Encourage some discussion of what the blue ring will do. Ask students to explain the phrase "doubles your age."

Ask the class to look at the table in problem 1. Invite students to look at the first line of the top table and discuss what it shows. Explain that their job is to fill in all the empty boxes in both tables. At first they will pretend to be 10 years old. In the second table, where the first column is blank, they may choose to be any age. They may also choose what rings they will put on.

Now invite students to look at problems 3 through 7. If you think students would benefit from hearing the problems, read them aloud to the class. They will record their answer for each problem in the space under that problem. Tell them that it is important to explain their answers clearly on all of these problems.

Arrange for students to work in pairs. These should be different pairs from those for any other assessment tasks being given at the same time. Each student should complete the written work individually.

Time allotment: **Before** ≈ 10 minutes **During** ≈ 35 minutes **After** ≈ X minutes.

Notes
"materials: calculator"
possible hints:

key ideas for discussion:

conducting discussion:

before:

after:

specific groups or children from whom to gather data:
This task is "ramped" so that virtually every student will be able to do something with the first question, but the challenge increases with each subsequent question.

what to do with written work:

Linking to Key Teaching Strategies

Mathematics Environment
As a problem task this assessment activity is reflective of a problem- solving environment.

Worthwhile Mathematical Task
"You may wish to extend this task by using some of the ideas that follow.

- Make up your own question about these magic rings. Work out the answer. Ask someone else to answer your question.
- Are there any ages that are impossible to produce with the rings?
- Invent different types of rings that also change your age. What would be the most useful set of rings to have?
- Investigate what happens when you put 2 rings in a different order (blue first, then green OR green first, then blue). For which combinations of rings does order matter? Are there combinations in which order doesn't make any difference? Why is this?" (*Balanced Assessment for the Mathematics Curriculum, 2000,*)

Cooperative Learning
Although designed to assess individual reporting, the fact that it encourages initial discussion in pairs acknowledges the power of group learning.

Models
This task encourages the development of symbolic representations of relationships with the abstraction of "age." This can be facilitated by the use of different models to reinforce meaning . E.g., counting blocks used to represent an age can be doubled or some "added" or "taken away."

Discourse, Writing and Active Listening
The nature of the student response activity as well as the before activity, has potential to engage the learners in both meaningful discourse and writing.

The request in questions 3 - 7 to "Explain your answer" is intended to encourage learners to justify their answer. Depending upon the responses, you may have to think about how to prepare learners to understand what this means. Many learners (without such preparation) often respond with a procedural description rather than a conceptual one.

Magic Age Rings

Name: _____

You are ten years old and as a birthday present you have
been given three boxes of magic rings

Caution—Only wear on the small finger of your left hand. **BLUE RINGS** THIS RING DOUBLES YOUR AGE.	Caution—Only wear on the small finger of your left hand. **GREEN RINGS** THIS RING ADDS 5 YEARS TO YOUR AGE.	Caution—Only wear on the small finger of your left hand. **YELLOW RINGS** THIS RING TAKES 2 YEARS AWAY FROM YOUR AGE.

With great care you slip one of the green rings onto the little finger of your left
hand. At once you start to grow and, within seconds, the magic has
worked— you are 15 years old. Discuss this with your partner. What does it
mean to "double" your age?

1. Complete the table.

Imagine that you are …	Rings you wear	Age you become	Number sentences to explain this
10 years old	Green	15	10 + 5 = 15
10 years old	Blue		
10 years old	Yellow		
10 years old	Green and then Blue	30	10 + 5 = 15 15 + 15 = 30
10 years old	Green and then Yellow	13	10 + 5 = 15 15 - 2 = 13
10 years old	Green and another Green		
10 years old	Yellow and then Blue		
10 years old	Yellow and another Yellow		

2. Use your imagination. Choose to be any age and wear any combination of rings. Complete the table.

Imagine that you are …	Rings you wear	Age you become	Number sentences to explain this

3. Latosha is 11 years old. She puts on three blue rings. How old is she now? Explain your answer.

4. Margaux is 10 years old but she wants to be 18. What rings should she put on to change her age to 18 years? Explain your answer.

5. Lloyd is 9 years old. He has a blue ring and a green ring. He wants to wear both rings. Will it make any difference to his age which ring he puts on first? Explain your answer.

6. Jessica is wearing a yellow ring and she tells you she is 13 years old. She takes off the yellow ring and she is back to her real age. What is her real age? Explain your answer.

7. Rashad is wearing three rings and is 12 years old. First he takes off a yellow ring, then a blue ring, and finally a green ring. He is back to his real age. How old is he? Explain your answer.

Magic Age Rings: Sample Solution

	Imagine that you are …	Rings you wear	Age you become	Number sentences to explain this
1)	10 years old	Green	15	10 + 5 = 15
	10 years old	Blue	20	10 + 10 = 20
	10 years old	Yellow	8	10 - 2 = 8
	10 years old	Green and then Blue	30	10 + 5 = 15, 15 + 15 = 30
	10 years old	Green and then Yellow	13	10 + 5 = 15, 15 - 2 = 13
	10 years old	Green and another Green	20	10 + 5 = 15, 15 + 5 = 20
	10 years old	Yellow and then Blue	16	10 - 2 = 8, 8 + 8 = 16
	10 years old	Yellow and another Yellow	6	10 - 2 = 8, 8 - 2 = 6

* Other number sentences may also be correct. For example, doubling may be shown as "× 2."

2) There is no reason why students should not complete this table with examples that use three or more rings— for example, "Blue, then Green, then Yellow."

3) Latosha will be 88 years old. $11 \times 2 = 22, 22 \times 2 = 44, 44 \times 2 = 88$

4) Margaux could put on a blue ring and then a yellow ring. $10 \times 2 = 20, 20 - 2 = 18$. Or she could put on two green rings and then a yellow ring. $10 + 5 = 15, 15 + 5 = 20, 20 - 2 = 18$. There are other solutions

5) It does make a difference which ring Lloyd puts on first. If he puts on the blue ring first, his age becomes 23 years. $9 \times 2 = 18, 18 + 5 = 23$ But if he puts on the green ring first, he becomes 28 years old. $9 + 5 = 14, 14 \times 2 = 28$

6) Jessica's real age is 15 years. A yellow ring takes 2 years off your age. So removing a yellow ring will do the opposite. It will make Jessica two years older. $13 + 2 = 15$

7) Rashad's real age is 2 years. $12 + 2 = 14, 14 \div 2 = 7, 7 - 5 = 2$. This last question is difficult as it really demands some understanding of inverse operations.

Rubric

Characterizing Performance

This section offers a characterization of student responses and provides indications of the ways in which the students were successful or unsuccessful in engaging with and completing the task. The descriptions are keyed to the *Core Elements of Performance.* Our global descriptions of student work range from "The student needs significant instruction" to "The student's work meets the essential demands of the task." The characterization of student responses for this task is based on these *Core Elements of Performance.*

1. Apply simple functions in a problem-solving situation.
2. Use combinations of various arithmetic operations to solve problems.
3. Demonstrate understanding of what effect reversing the order of operations has on a problem.
4. Use doubling and an understanding of inverse operations to solve complex problems.
5. Explain how answers are decided.

Descriptions of Student Work

The student needs significant instruction. These papers show evidence of some limited success in one or two of the core elements of performance, most commonly in the first and second.

The student needs some instruction. These papers provide evidence of ability in the first two core elements, both in the tables and in some of the problems. There may be some limited evidence of performance in the last three core elements: this evidence, however, is weak and inconsistent. This level of work will show the ability to apply simple functions and to combine various arithmetic operations not just in the table but also in some of the problems 3 through 7. Generally, there will not be evidence of an ability to work with inverse operations, or to double numbers correctly or consistently. The paper will not provide evidence of understanding the importance of order of operations. Explanations will be limited and/or weak.

The student's work needs to be revised. There will be evidence of ability to perform in at least four out of five of the core elements of performance. There may be some inconsistency in one or two core elements (for example, the response may correctly show doubling in the table, but incorrectly solve problem 3, or correctly solve problem 5, but make a mistake in order of operations in the table). The answers may be all correct, but missing any explanation.

The student's work meets the essential demands of the task.
There will be no mistakes (or only very minor mistakes) in the tables. Problems 3 through 7 will fully demonstrate ability in all five core elements of performance. There may be an error in one of the elements of performance; however, that element will be correctly demonstrated elsewhere. (For example, many otherwise very strong responses will make a mistake on problem 7.)

This task comes from **Elementary Grades Assessment Package 1, Dale Seymour Publications, 2000 (ISBN: 0-7690-0063-0). The task and supporting materials were developed by the project Balanced Assessment for the Mathematics Curriculum. Further information about additional packages may be obtained from the publisher. Further information about additional tasks and supporting instructional and professional development materials may be obtained from the Balanced Assessment-Mathematics Assessment Resource Service project website: http://www.educ.msu.edu/mars/*

30. Grocery Store

Source: Balanced Assessment for the Mathematics Curriculum*

Grade: 5 - 8 (can be modified to address what you know about your students).

Math: I want the learners to: Reason using ratio and proportion. Reason algebraically. Generalize symbolically.

Task: This task asks students to consider the planning of a layout for a new grocery store. Students answer questions using scale models of shopping carts to solve problems related to the store's floor plan.

Expectation: It is assumed that students have had experience with ratio and proportion and with generalizing linear situations symbolically...Students may discuss the task in pairs, but then complete an individual written response.

BEFORE activity: Read through the task. Students should be familiar with the context of this task. If not, have the class discuss how shopping carts are placed in a grocery store and what "nested" means.

Time allotment: Before ≈ 10 minutes During ≈ 35 minutes After ≈ X minutes

---

Notes
"materials: rulers and calculators"
possible hints:

key ideas for discussion:

conducting discussion:

before:

after:Several questions related to accuracy could arise and you may want to discuss them, with your class after students have completed the task. For example, how much should we rely on the accuracy of the scale model provided? What happens if we assume that it could be off by 5%? Is the 10 meter space a corral by the entrance of the store and is it fine if the carts stick out a bit? Or is the 10 meter space the width of a store room in which the carts must fit? How would Rasheed's purpose affect our answers? What other purposes might make us respond differently?

specific groups or children from whom to gather data:
This task is "ramped" so that virtually every student will be able to do something with the first question, but the challenge increases with each subsequent question

what to do with written work:

Name: _____

Grocery Store

Rasheed is planning the layout for a new grocery store. He found the diagram below in a supply catalog. It shows a drawing of a single shopping cart and a drawing of 12 shopping carts that are "nested" together. (The drawings are $\frac{1}{24}$ th of the real size.)

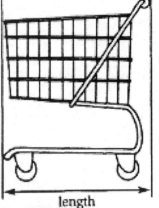

length

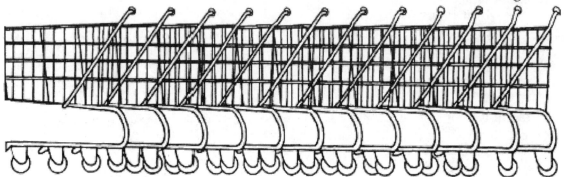

Rasheed has several questions:

1. What is the length of a real shopping cart?

2. When the real carts are nested, how much does each cart stick out beyond the next one in the line?

3. What would be the total length of a row of 20 real nested carts?

4. What rule or formula could I use to find the length of a row of real nested carts for any (*n*) number of carts?

5. How many real nested carts could fit in a space 10 meters long?

Write a letter to Rasheed that answers his questions.
- For each question, explain your answer so that he can understand it and use it to make decisions about the store.
- To explain question 4, you may want to draw and label a diagram that tells what each part of your formula represents.

Grocery Store: Sample Solution

The following is a sample solution using centimeters. Students may also use inch measurements, although this will require conversion to metric for the last question.

Dear Rasheed,

Here are answers to your questions. I hope they will help you in planning your store. Good luck!

My answer to question 1— The length of a real shopping cart is 96 cm. or .96 m. The scale model of a single shopping cart measures 4 cm. Since the scale is 1 to 24, multiply by 24 to get the real length: 4 cm × 24 = 96 cm.

My answer to question 2— Each cart sticks out beyond the next one in line approximately 26.4 cm or about .26 m. In the scale model of the nested carts, I measured the distance between the handles on the first and second shopping carts and got about 1.1 cm (although it looks like it's a little more than that). You could measure in other ways too— by the grills, between the last two carts, and so on. Anyway, it comes out roughly the same. Then I multiplied by 24 to get the real length: 1.1 cm × 24 = 26.4 cm.

My answer to question 3— The total length of 20 nested carts is approximately 5.98 m. In any row of nested carts, the first cart will take up 96 cm. Each additional cart will add 26.4 cm to the total length. Since you need the length of 20 carts, I added the length of one cart and 19 "stick outs" to get 96 cm + (19 × 26.4) = 597.6 cm or about 5.98 m.

My answer to question 4— If L is the total length in centimeters of n nested carts, then the total length of n nested carts is $L = 96 + 26.4(n - 1)$ which can be written as: $L = 26.4n + 69.6$. I got this formula in a way similar to finding the answer to your last question. The 96 is for the length of the first cart. After the first cart, there are $n - 1$ carts sticking out. So I multiplied the number of "stick outs" $(n - 1)$ by the length of each "stick out" (26.4) and added it all to the length of the first cart (96) giving me $L = 96 + 26.4(n - 1)$. Remember that this formula gives length in centimeters.

My answer to question 5— About 35 carts could fit in a space 10 meters long. I got this answer by using the above formula and letting $L = 10$m. Since the formula finds length in centimeters, I let $L = 1000$ cm. Plugging in I got 1000 - 26.4n + 69.6. I solved for n: $n = 35.24$. Since you can't have parts of shopping carts, I rounded this number down to 35.

I hope that these answers help you to make decisions about your store.

Sincerely,
A. Student

Rubric

Characterizing Performance

This section offers a characterization of student responses and provides indications of the ways in which the students were successful or unsuccessful in engaging with and completing the task. The descriptions are keyed to the *Core Elements of Performance*. Our global descriptions of student work range from "The student needs significant instruction" to "The student's work meets the essential demands of the task."

The characterization of student responses for this task is based on these *Core Elements of Performance*:
1. Reason using ratio and proportion and successfully move back and forth between a real work situation and a scale model to determine the lengths of shopping carts
2. Use algebraic reasoning to solve for an unknown and to generalize a linear relationship symbolically
3. Communicate mathematical reasoning

Descriptions of Student Work

The student needs significant instruction. Student may answer some questions correctly (either in the letter to Rasheed or beside the question), but does not successfully find an unknown (as in questions 3 and 5) or formulate a symbolic generalization (as in question 4).

The student needs some instruction. Student successfully finds an unknown (questions 3 and 5), but does not demonstrate an understanding of how to arrive at a general formula

The student's work needs to be revised. Student successfully finds the unknown (questions 3 and 5), *AND* shows an understanding of how to generalize the situation, but fails to arrive at a completely correct formula.

The student's work meets the essential demands of the task. Student successfully finds the unknown and arrives at a correct generalization of the situation. Minor errors that do not distort the reasonableness of solutions are permitted.

This task comes from **Middle Grades Assessment Package 2, Dale Seymour Publications, 2000 (ISBN: 0-7690-0067-3). The task and supporting materials were developed by the project Balanced Assessment for the Mathematics Curriculum. Further information about additional packages may be obtained from the publisher. Further information about additional tasks and supporting instructional and professional development materials may be obtained from the Balanced Assessment-Mathematics Assessment Resource Service project website: http://www.educ.msu.edu/mars/*

● Field Experience Guide: Resources for Teachers of Elementary and Middle School Mathematics ● © Allyn & Bacon 2004

⋈   Creating Your Own Manipulatives

Here are a number of manipulative models you can make to use as in class.

Create your own base-10 set!

Materials:
- Photo-copy machine that does enlarging
- Lamination facility (or access to card stock for copying)
- Square graph paper (see black line masters)
- Scissors or paper cutter

What to do:
1) Photo-copy graph paper to enlarge the squares to at least one-half inch by one-half inch (large enough for little hands to manipulate).
2) Make enough copies for each child to cut out a sufficient supply of "flats" (100 squares), "rods" (10 strips), and "singles" (1 single square).
3) Laminate the paper before cutting (or use card stock) for durability.
4) As an option, you could have children color the pieces.

Create your own 2-color counters!

Materials:
- Any dry beans from your local grocer (larger beans such as kidney beans or great northern beans work best).
- Can of non-toxic spray paint from your local hardware store.
- Drop cloth (or newspapers to catch the excess spray paint).

What to do:
1) Spread the drop cloth or newspaper.
2) Spread the beans into a single layer on the cloth (or newspaper).
3) Spray paint the beans.
4) Let dry.

Create your own centimeter rods!

Materials:
- Any computer drawing program.
- Color printer.
- Laminating facility (or white card stock).
- Scissors or paper cutters.

What to do:

Use your draw program to create colored rectangles that are 1 cm wide and adhere to the following pattern:

White	1 cm long
Red	2 cm long
Lt green	3 cm long
Purple	4 cm long
Yellow	5 cm long
Dk green	6 cm long
Black	7 cm long
Brown	8 cm long
Blue	9 cm long
Orange	10 cm long

Although these are the colors used by the Cuisenaire ® Rods, you could choose different colors. The key is consistency.

Print these in color in sufficient numbers to allow each child to have some of each.

Laminate them before cutting (or use card stock) to enhance durability.

Ellison® Die Cuts

If your school or teacher resource center has an Ellison Die cutter, you can create nearly any manipulative for which dies exist. This machine can be a cost effective way of securing manipulatives such as attribute blocks, pattern blocks, circular fraction sets, and counters. These machines use stencil like material called *dies* to cut textile or paper into specific shapes.

For additional ideas, visit Ellison's web site at http://www.ellison.com

Create your own tangrams!

Material:
> Regular 8.5" x 11" paper

What to do:

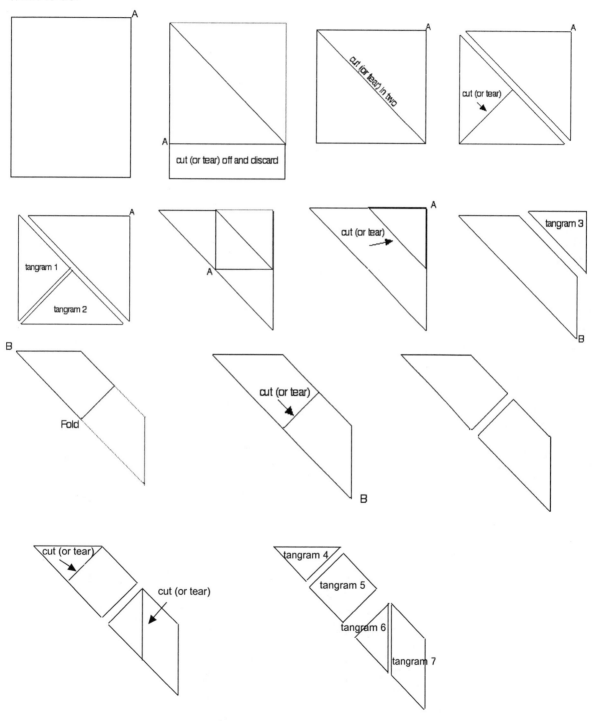

4. Blackline Masters

The reproducible blackline masters featured in this section can be used for making instructional materials to support activities.

Card Stock Materials

A good way to have many materials made quickly and easily for students is to have them duplicated on card stock at a photo copy store. Card stock is a heavy paper that comes in a variety of colors. It is also called *cover stock* or *index stock*. The price is about twice that of paper.

Card stock can be laminated and then cut into smaller pieces, if desired. The laminate adheres very well. Laminate first, and then cut into pieces afterward. Otherwise you will need to cut each piece twice.

Materials are best kept in plastic bags with zip-type closures. Freezer bags are recommended for durability. Punch a hole near the top of the bag so that you do not store air. Lots of small bags can be stuffed into the largest bags. You can always see what you have stored in the bags.

The following list is a suggestion for materials that can be made from card stock using the masters in this section. Quantity suggestions are also given.

Dot Cards

One complete set of cards will serve four to six children. Duplicate each set in a different color so that mixed sets can be separated easily. Laminate and then cut with a paper cutter.

Five-Frames and Ten-Frames

Five-frames and ten-frames are best duplicated on light- colored card stock. Do not laminate; if you do, the mats will curl and counters will slide around.

10 x 10 Multiplication Array

Make one per student in any color. Lamination is suggested. Provide each student with an L-shaped piece of tagboard.

Base-Ten Pieces (Centimeter Grid)

Use the grid (number 11) to make a master as directed. Run copies on white card stock. One sheet will make 4 hundreds and 10 tens or 4 hundreds and a lot of ones. Mount the printed card stock on white poster board using either a dry-mount press or permanent spray adhesive. (Spray adhesive can be purchased in art supply stores. It is very effective but messy to handle.) Cut into pieces with a paper cutter. For the tens and ones pieces, it is recommended that you mount the index stock onto *mount board* or *illustration board*, also available in art supply stores. This material is thicker and will make the pieces easier to handle. It is recommended that you *not* laminate the base-ten pieces. A kit consisting of 10 hundreds, 30 tens, and 30 ones is adequate for each student or pair of students.

Bean Stick Base-Ten Pieces

Use dried beans (great northern, pinto, etc.) to make bean sticks. Craft sticks can be purchased in craft stores in boxes of 500. Use white glue (such as Elmer's) or a glue gun. Also dribble a row of glue over the beans to keep them from splitting off the sticks. (The white glue dries clear.) For hundreds, use the master mounted on poster board.

Little Ten-Frames

There are two masters for these materials. One has full ten-frames and the other has 1 to 9 dots, including two with 5 dots. Copy the 1-to-9 master on one color of card stock and the full ten-frames on another. Cut off most of the excess stock (do not trim) and then laminate. Cut into little ten-frames. Each set consists of 20 pieces: 10 full ten-frames and 10 of the 1-to-9 pieces, including 2 fives. Make a set for each child.

Place-Value Mat (with Ten-Frames)

Mats can be duplicated on any pastel card stock. It is recommended that you not laminate these because they tend to curl and counters slide around too much. Make one for every child.

Circular Fraction Pieces

First make three copies of each page of the master. Cut the disks apart and tape onto blank pages with three of the same type on a page. You will then have a separate master for each size with three full circles per master. Duplicate each master on a different color card stock. Laminate and then cut the circles out. A kit for one or two students should have two circles of each size piece.

Hundredths Disk

These disks can be made on paper but are much more satisfying on card stock. Duplicate the master on two contrasting colors. Laminate and cut the circles and also the slot on the dotted line. Make a set for each student. It's easy and worthwhile.

Tangrams and Mosaic Puzzle

Both tangrams and the Mosaic Puzzle should be copied on card stock. For younger children, the card stock should first be mounted on poster board to make the pieces a bit thicker and easier to put together in puzzles. You will want one set of each per student.

Woozle Cards

Copy the Woozle Card master on white or off-white card stock. You need two copies per set. Before laminating, color one set one color and the other a different color. An easy way to color the cards is to make one pass around the inside of each Woozle, leaving the rest of the creature white. If you color the entire Woozle, the dots may not show up. Make one set for every four students.

Transparencies and Overhead Models

A copy of any page can be made into a transparency with a photocopier.

Some masters make fine transparency mats to use for demonstration purposes on the overhead. The 10 x 10 array, the blank hundreds board, and the large geoboard are examples. The five-frame and ten-frame work well with counters. The place-value mat can be used with strips and squares or with counters and cups directly on the overhead. The missing-part blank and the record blanks for the four algorithms are pages that you may wish to use as write-on transparencies.
A transparency of the 10,000 grid is the easiest way there is to show 10,000 or to model four-place decimal numbers.

A transparency of the degrees and wedges page is the very best way to illustrate what a degree is and also to help explain protractors.

All of the line and dot grids are useful to have available as transparencies. You may find it a good idea to make several copies of each and keep them in a folder where you can get to them easily. For the Woozle Cards, dot cards, little ten-frames, and assorted shapes, make a reduction of the master on a photocopy machine. Then make transparencies of the small cards, cut them apart, and use them on the overhead.

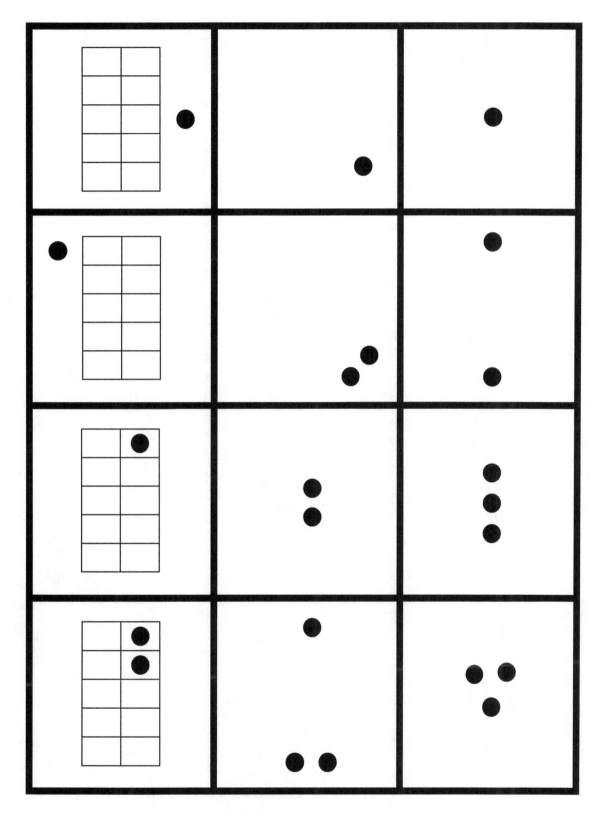

Dot cards—1

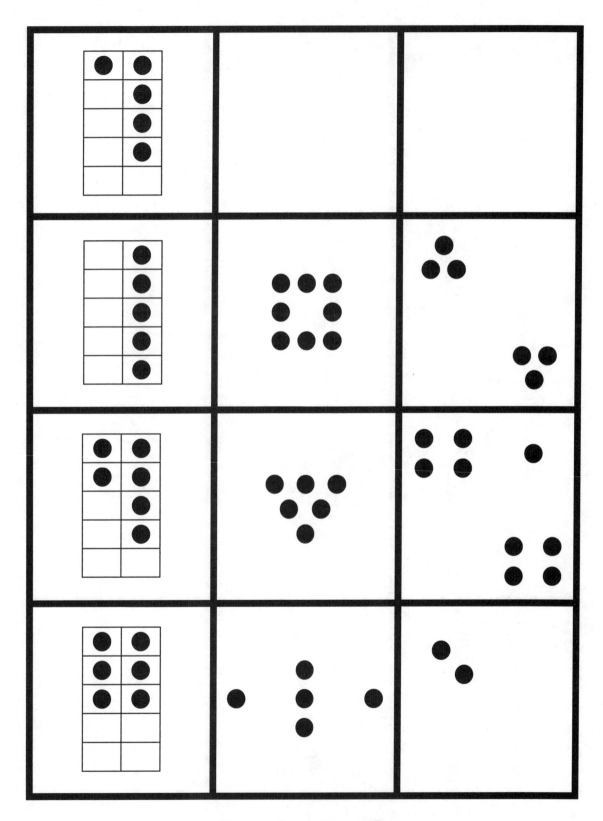

Dot cards—2

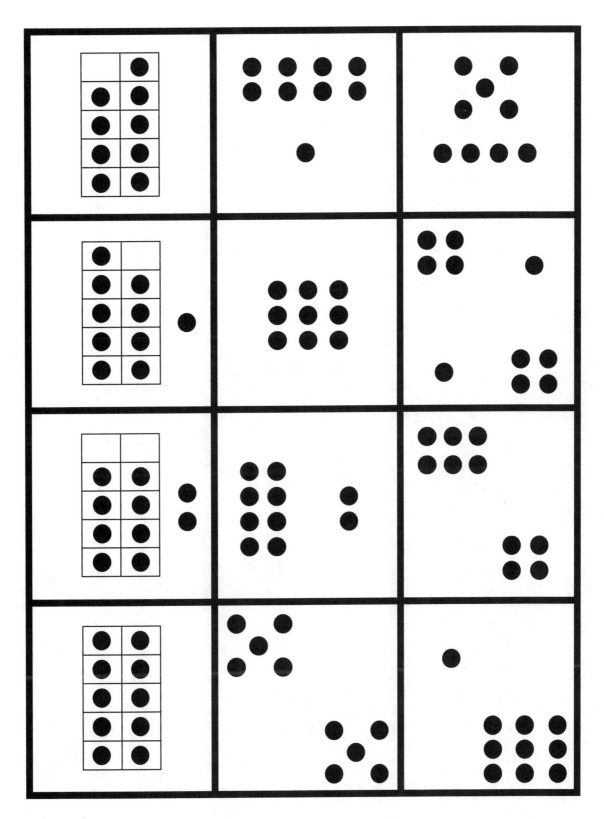

Dot cards—3

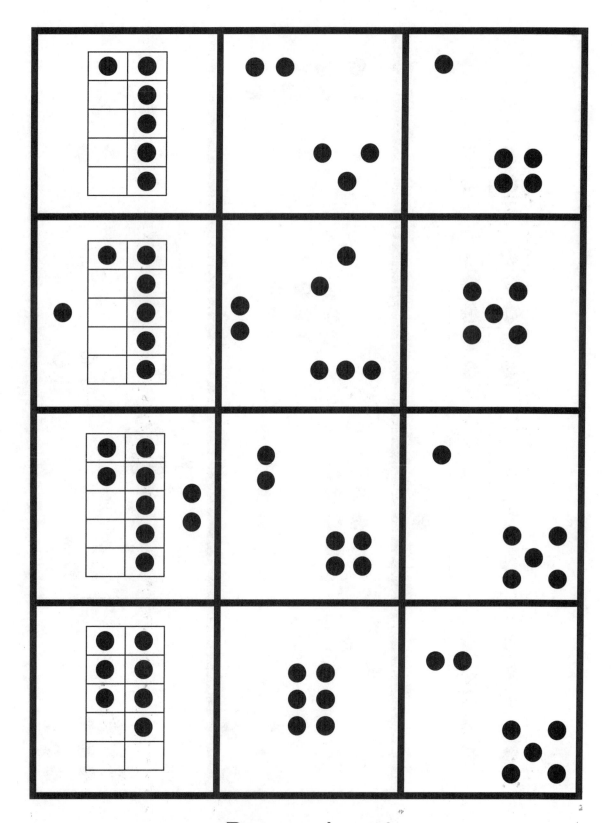

Dot cards—4

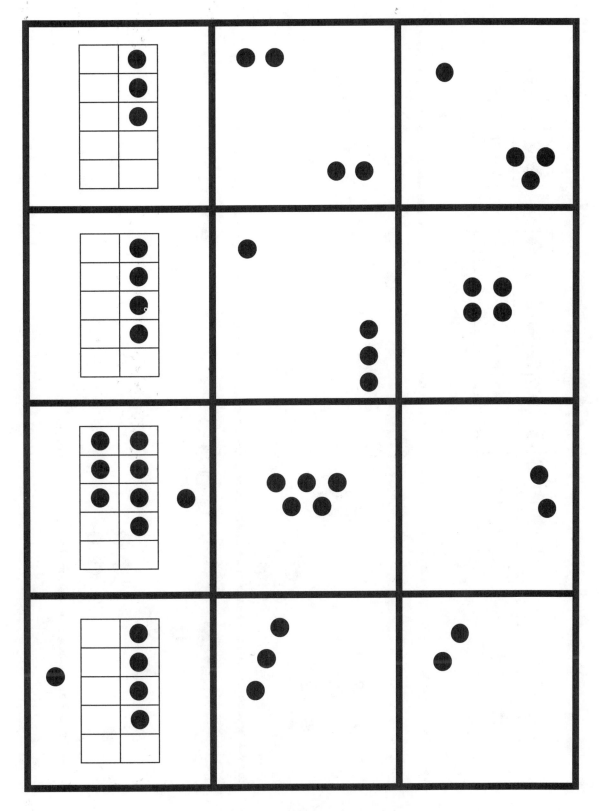

Dot cards—5

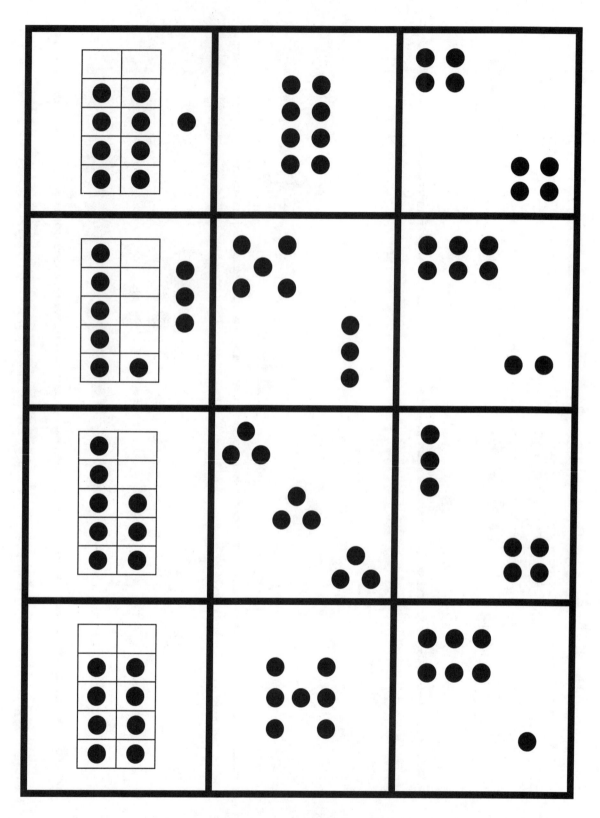

Dot cards—6

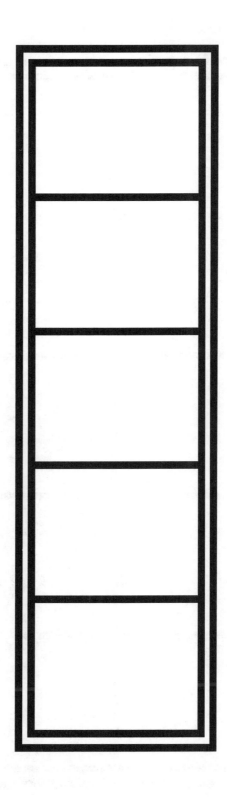

Five-frame — 7

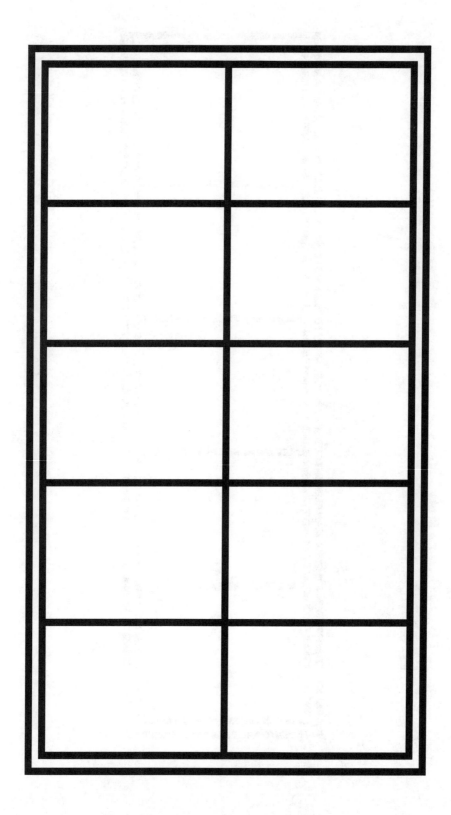

Ten-frame—8

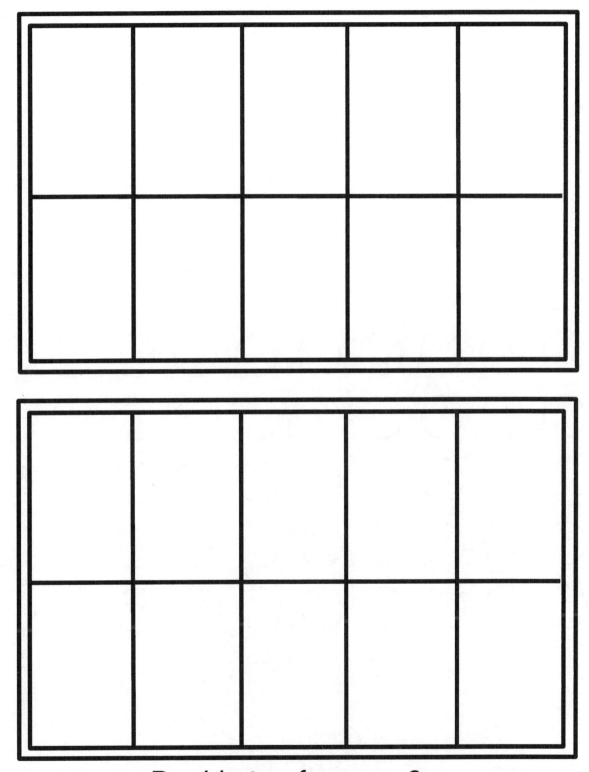

Double ten-frame — 9

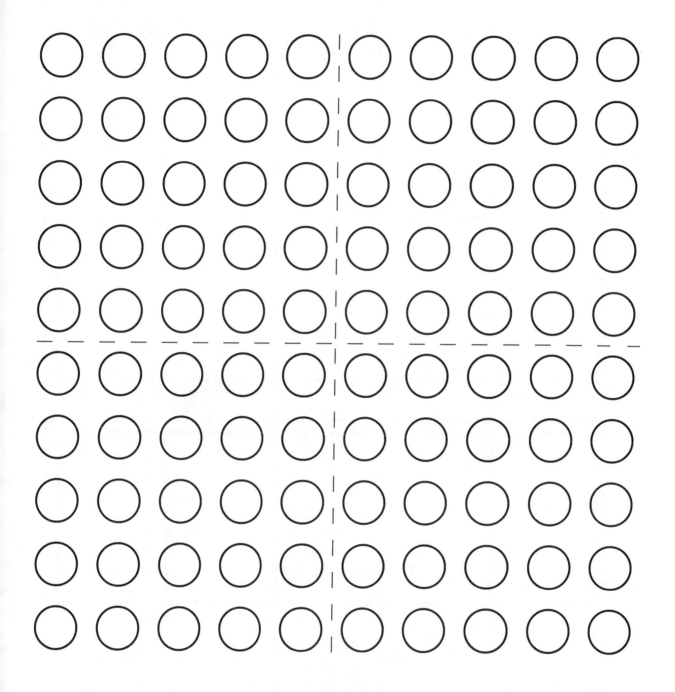

10 × 10 multiplication array—10

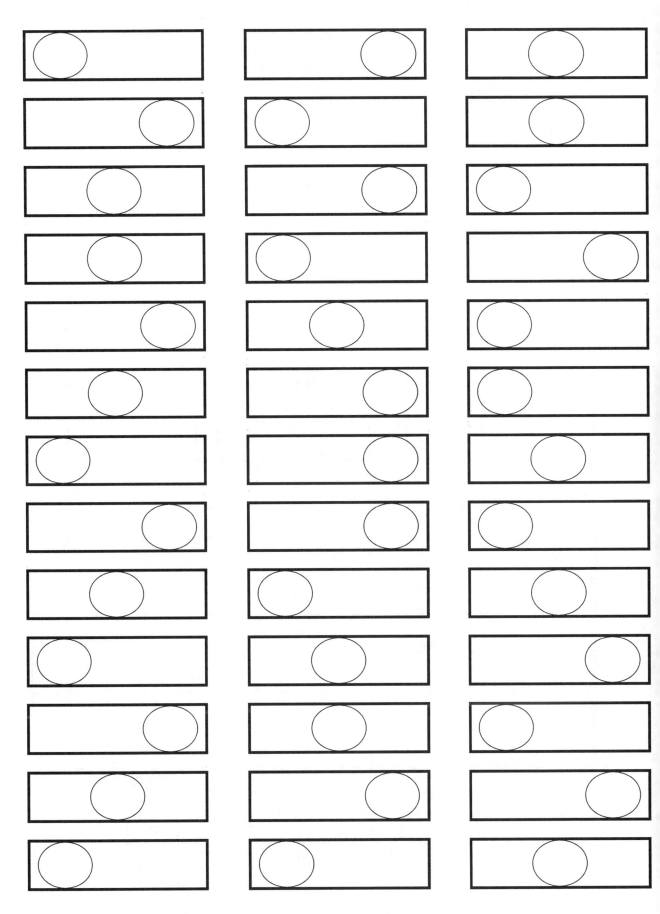

Missing-part blanks -11

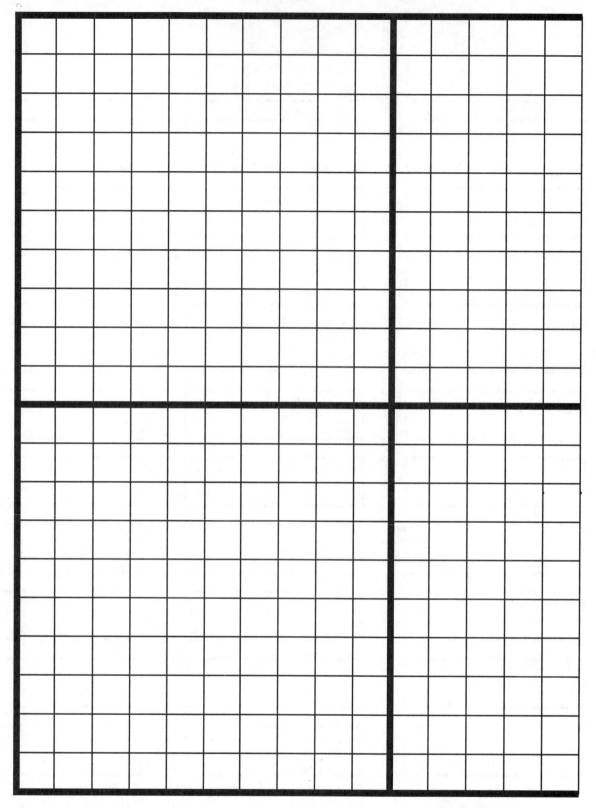

1. Make two copies of this page. Cut out the grid from each copy.
2. Overlap the two grids, and tape onto a blank sheet
   to form a 20-by-25-cm grid with 4 complete
   hundreds squares and 2 rows of 5 tens each.
3. Use this as a master to make copies on card stock.

Base-ten materials grid -12

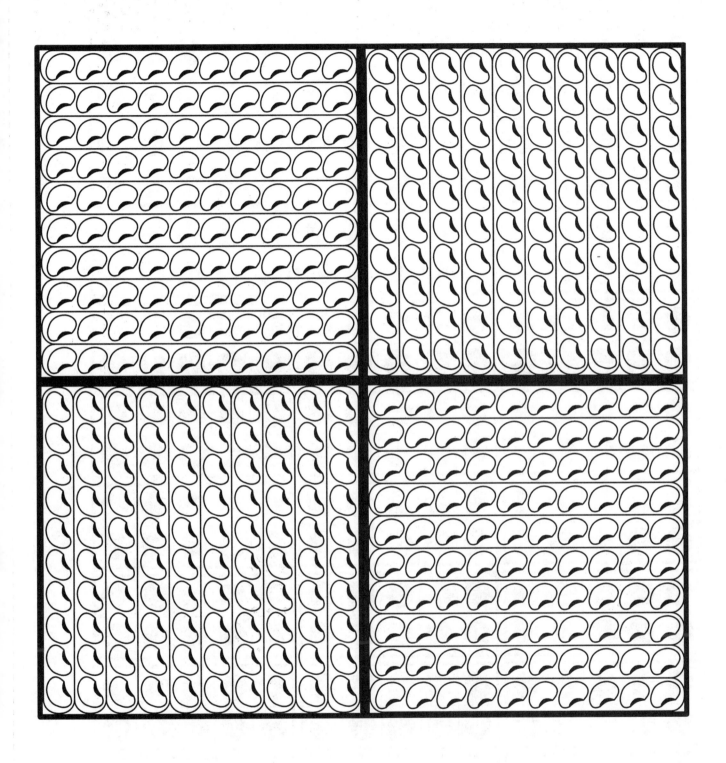

*Hundreds master for bean
stick base-ten pieces -13*

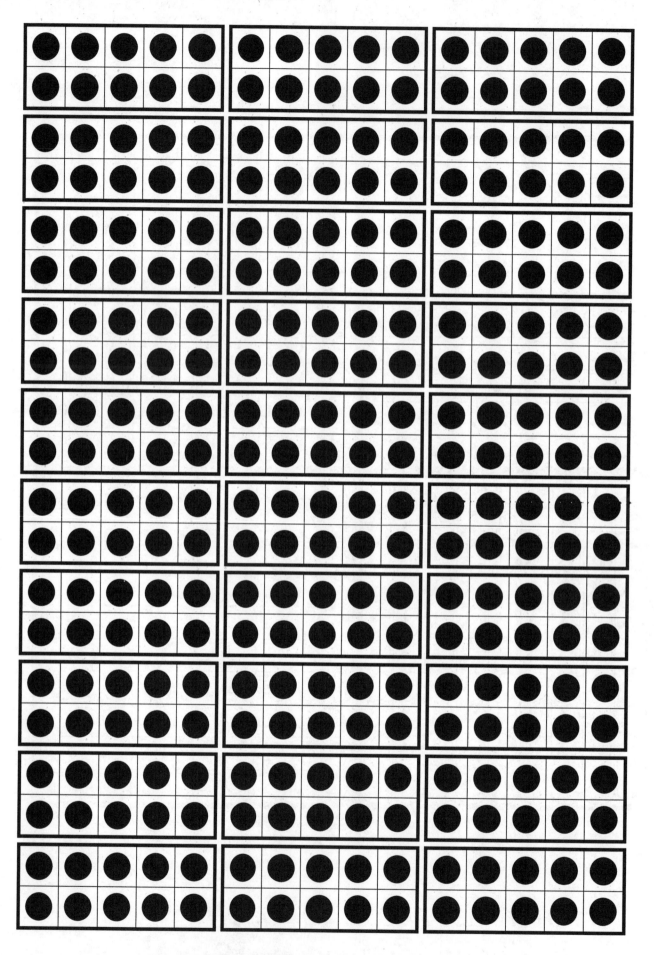

Little ten-frames -14

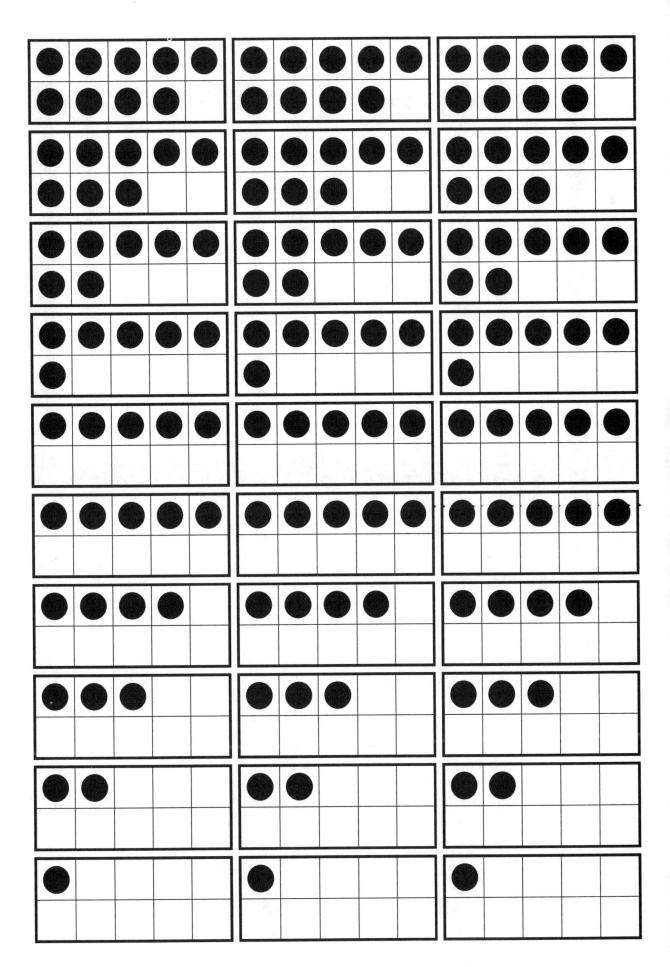

Little ten-frames -15

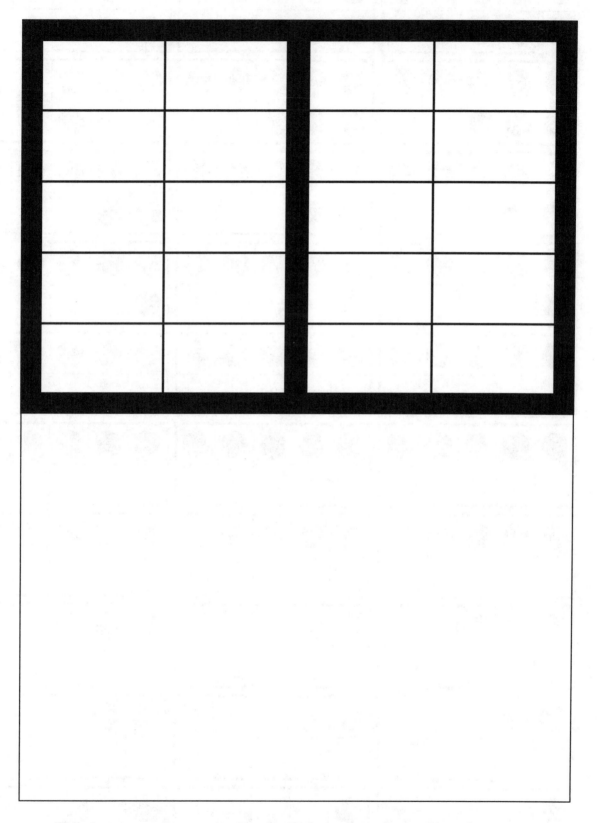

Place-value mat (with ten-frames)—16

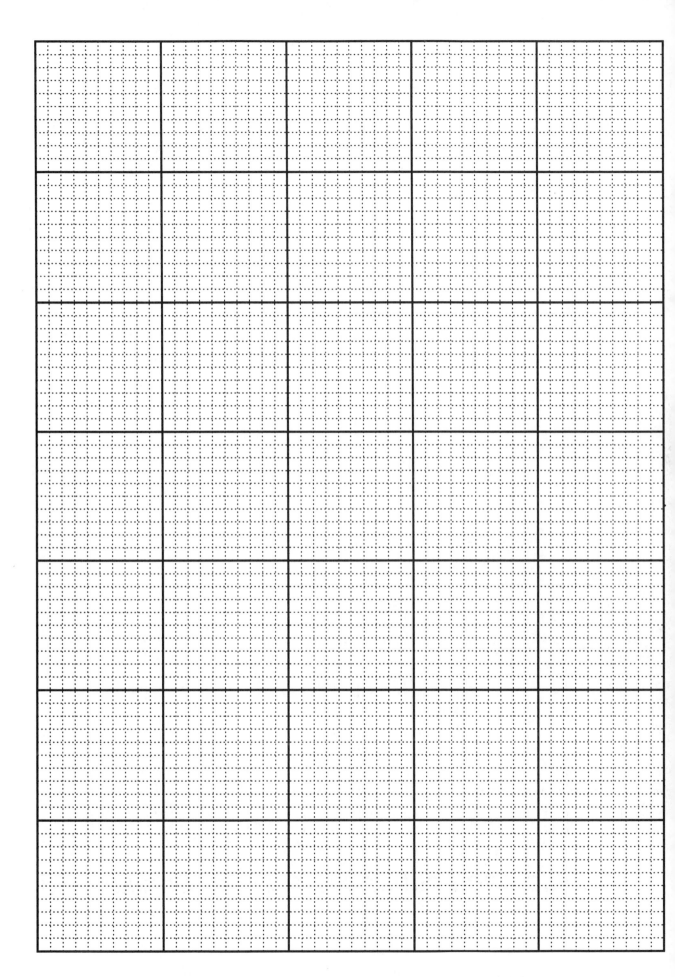

Base-ten grid paper -17

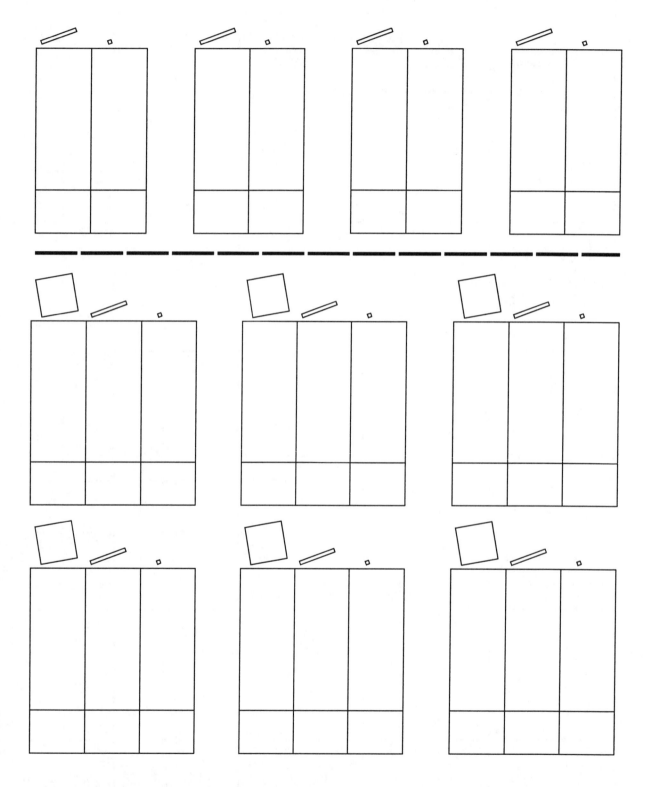

Addition and subtraction record blanks—18

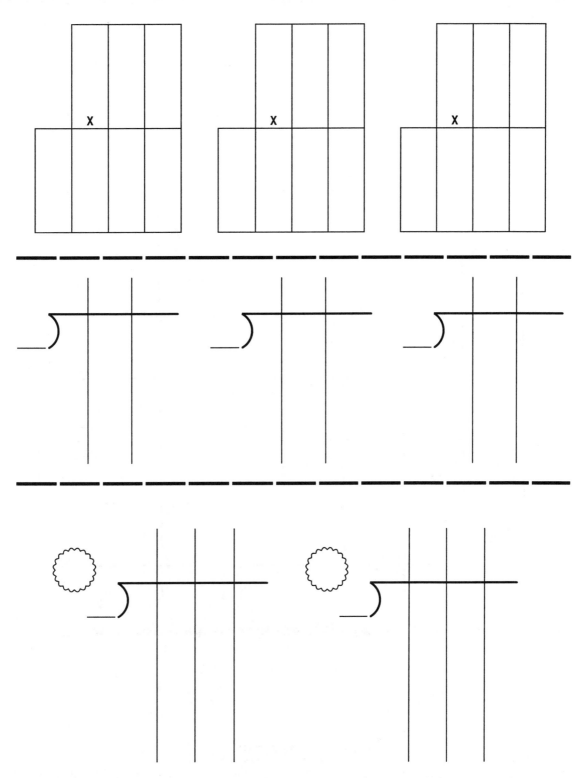

Multiplication and division record blanks—19

Blank hundreds chart (10 × 10 square) — 20

1	2	3	4	5	6	7	8	9	10
11	12	13	14	15	16	17	18	19	20
21	22	23	24	25	26	27	28	29	30
31	32	33	34	35	36	37	38	39	40
41	42	43	44	45	46	47	48	49	50
51	52	53	54	55	56	57	58	59	60
61	62	63	64	65	66	67	68	69	70
71	72	73	74	75	76	77	78	79	80
81	82	83	84	85	86	87	88	89	90
91	92	93	94	95	96	97	98	99	100

Hundreds chart—21

1	2	3	4	5	6	7	8	9	10
11	12	13	14	15	16	17	18	19	20
21	22	23	24	25	26	27	28	29	30
31	32	33	34	35	36	37	38	39	40
41	42	43	44	45	46	47	48	49	50
51	52	53	54	55	56	57	58	59	60
61	62	63	64	65	66	67	68	69	70
71	72	73	74	75	76	77	78	79	80
81	82	83	84	85	86	87	88	89	90
91	92	93	94	95	96	97	98	99	100

1	2	3	4	5	6	7	8	9	10
11	12	13	14	15	16	17	18	19	20
21	22	23	24	25	26	27	28	29	30
31	32	33	34	35	36	37	38	39	40
41	42	43	44	45	46	47	48	49	50
51	52	53	54	55	56	57	58	59	60
61	62	63	64	65	66	67	68	69	70
71	72	73	74	75	76	77	78	79	80
81	82	83	84	85	86	87	88	89	90
91	92	93	94	95	96	97	98	99	100

1	2	3	4	5	6	7	8	9	10
11	12	13	14	15	16	17	18	19	20
21	22	23	24	25	26	27	28	29	30
31	32	33	34	35	36	37	38	39	40
41	42	43	44	45	46	47	48	49	50
51	52	53	54	55	56	57	58	59	60
61	62	63	64	65	66	67	68	69	70
71	72	73	74	75	76	77	78	79	80
81	82	83	84	85	86	87	88	89	90
91	92	93	94	95	96	97	98	99	100

1	2	3	4	5	6	7	8	9	10
11	12	13	14	15	16	17	18	19	20
21	22	23	24	25	26	27	28	29	30
31	32	33	34	35	36	37	38	39	40
41	42	43	44	45	46	47	48	49	50
51	52	53	54	55	56	57	58	59	60
61	62	63	64	65	66	67	68	69	70
71	72	73	74	75	76	77	78	79	80
81	82	83	84	85	86	87	88	89	90
91	92	93	94	95	96	97	98	99	100

Four small hundreds charts—22

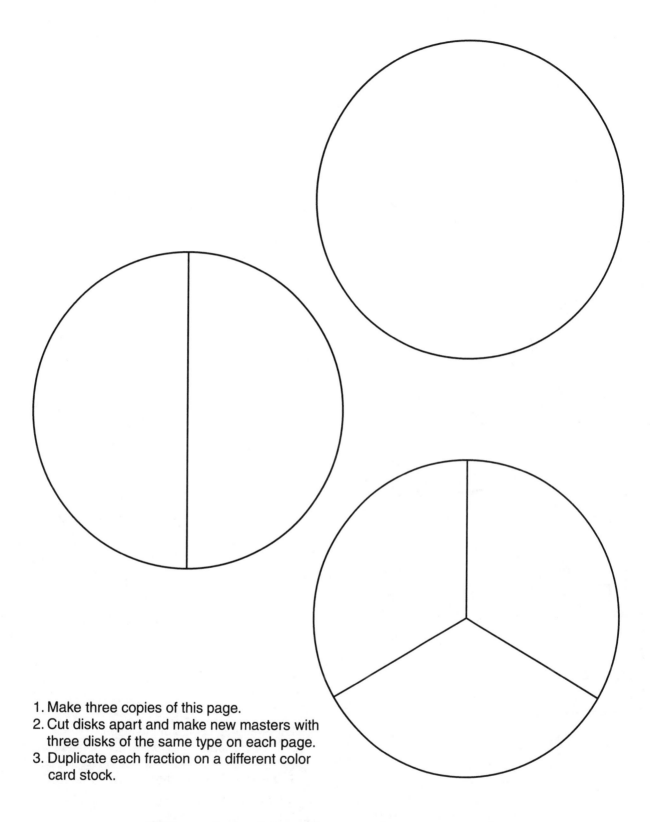

1. Make three copies of this page.
2. Cut disks apart and make new masters with three disks of the same type on each page.
3. Duplicate each fraction on a different color card stock.

Circular fraction pieces—23

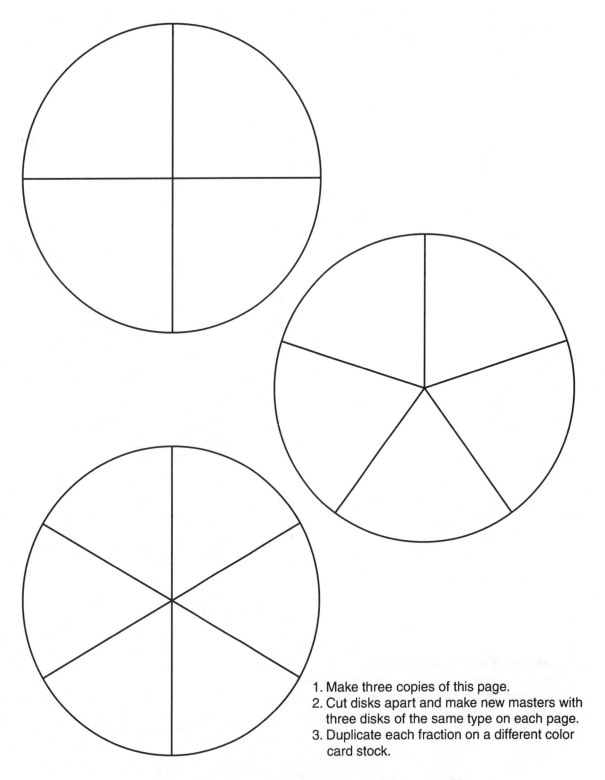

1. Make three copies of this page.
2. Cut disks apart and make new masters with three disks of the same type on each page.
3. Duplicate each fraction on a different color card stock.

Circular fraction pieces—24

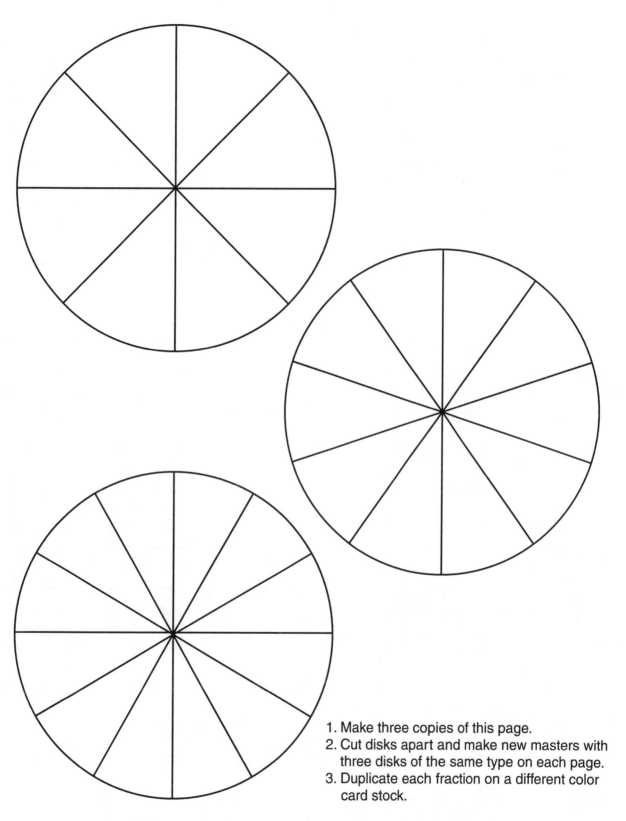

1. Make three copies of this page.
2. Cut disks apart and make new masters with three disks of the same type on each page.
3. Duplicate each fraction on a different color card stock.

Circular fraction pieces—25

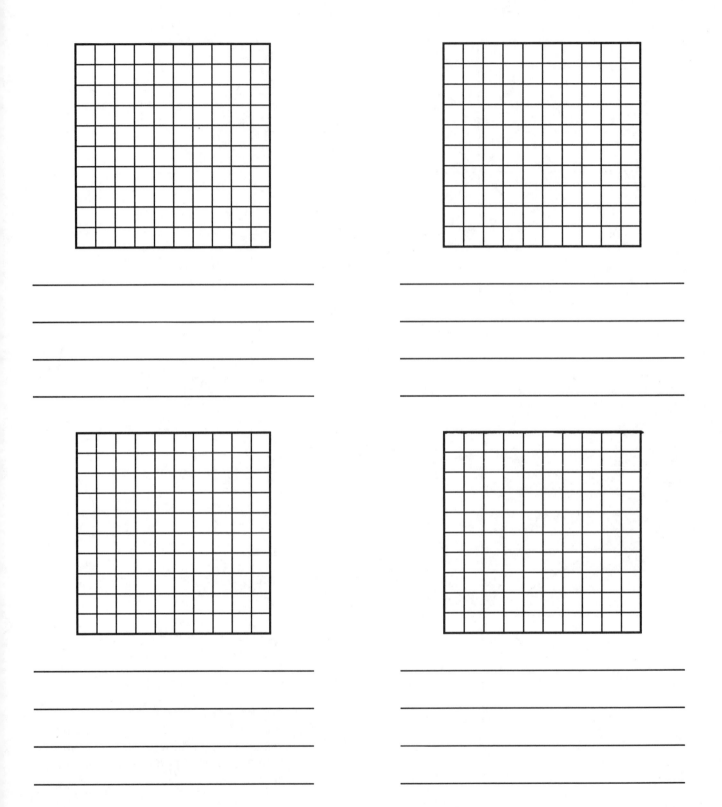

10 × 10 grids - 26

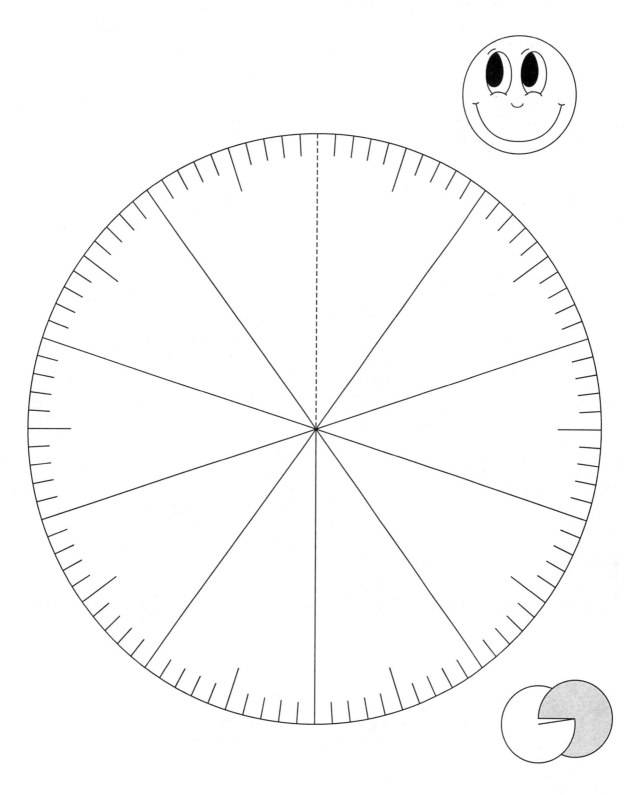

Hundredths disk—27

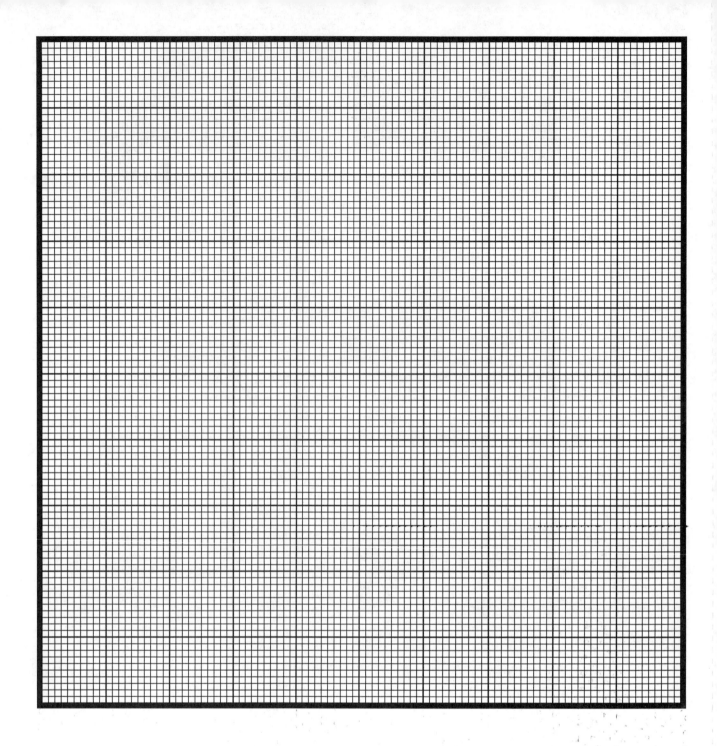

10,000 grid - 28

180 Degrees

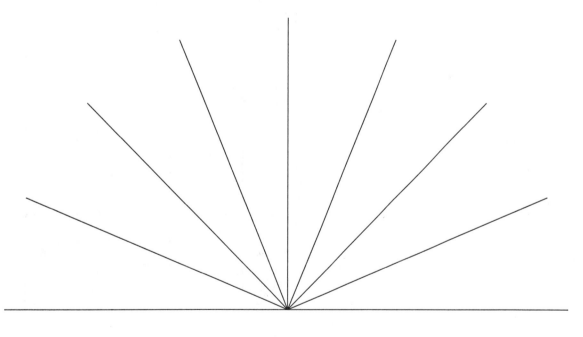

8 Wedges

Degrees and wedges—29

2-cm square grid - 30

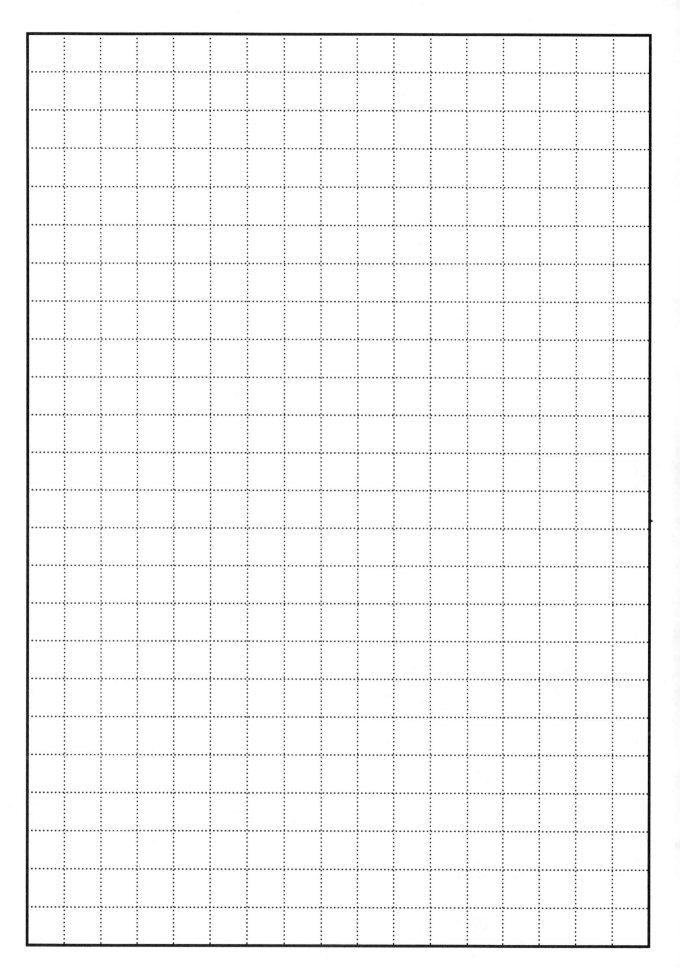

1-cm square grid - 31

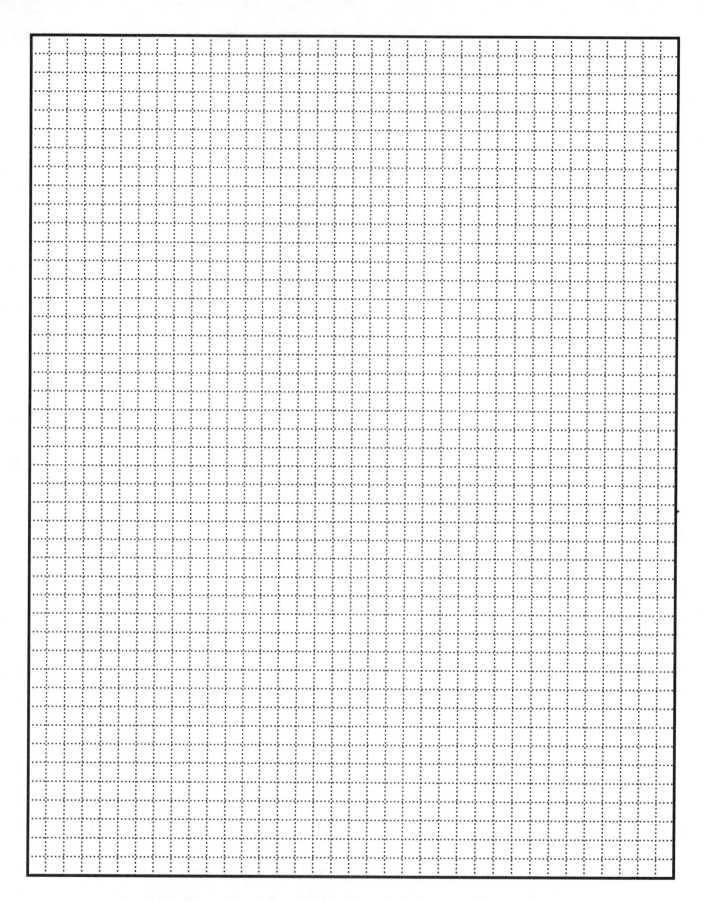

0.5-cm square grid - 32

1-cm square dot grid - 33

2-cm isometric grid - 34

1-cm isometric dot grid - 35

1-cm square/diagonal grid - 36

Assorted shapes—37 through 43

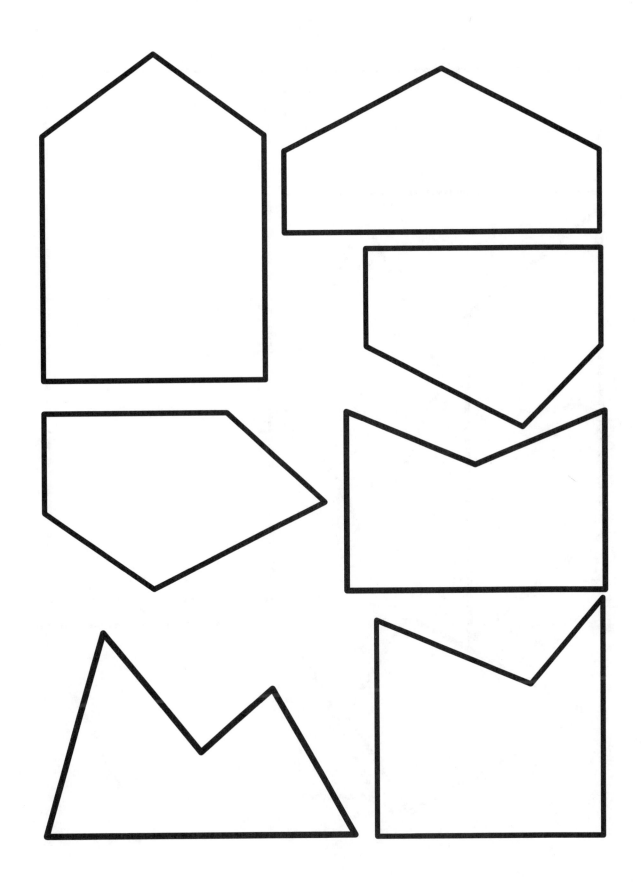

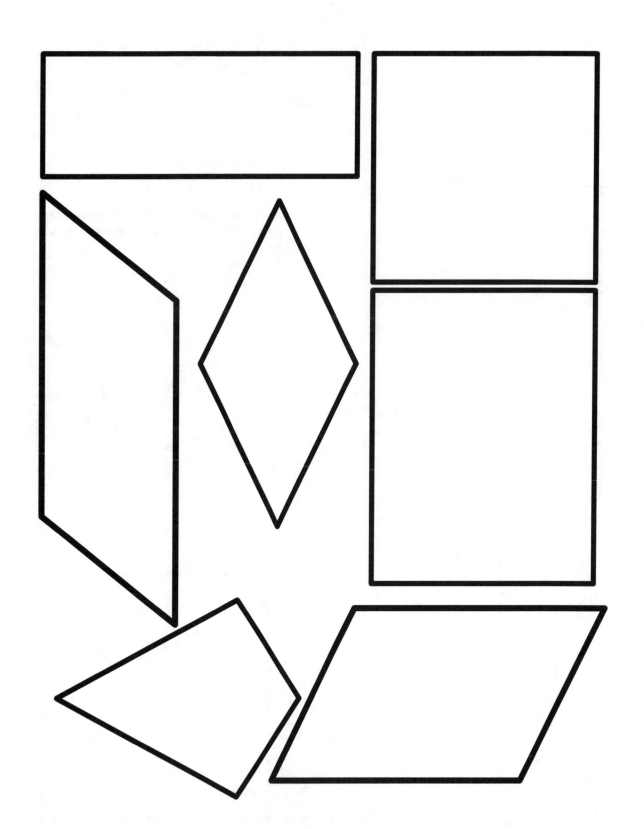

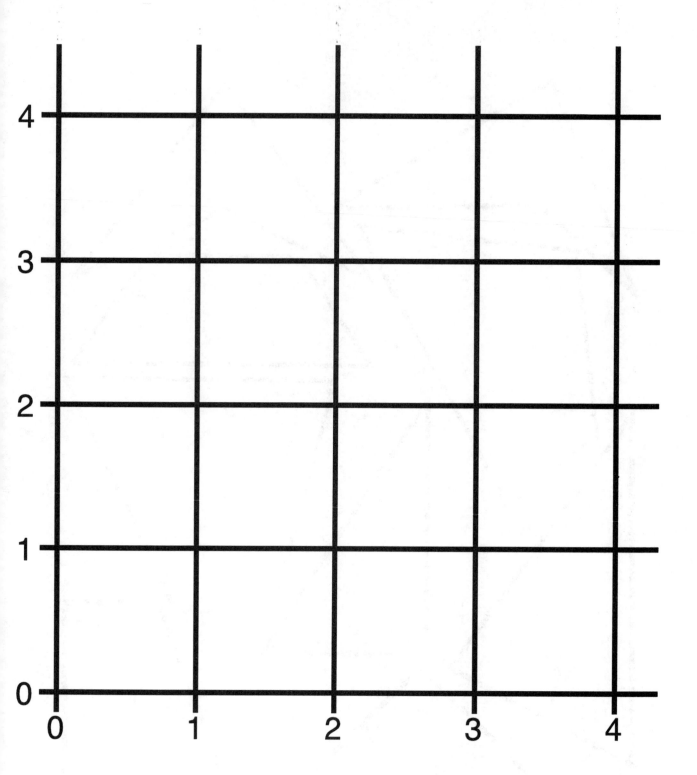

Coordinate grid - 44

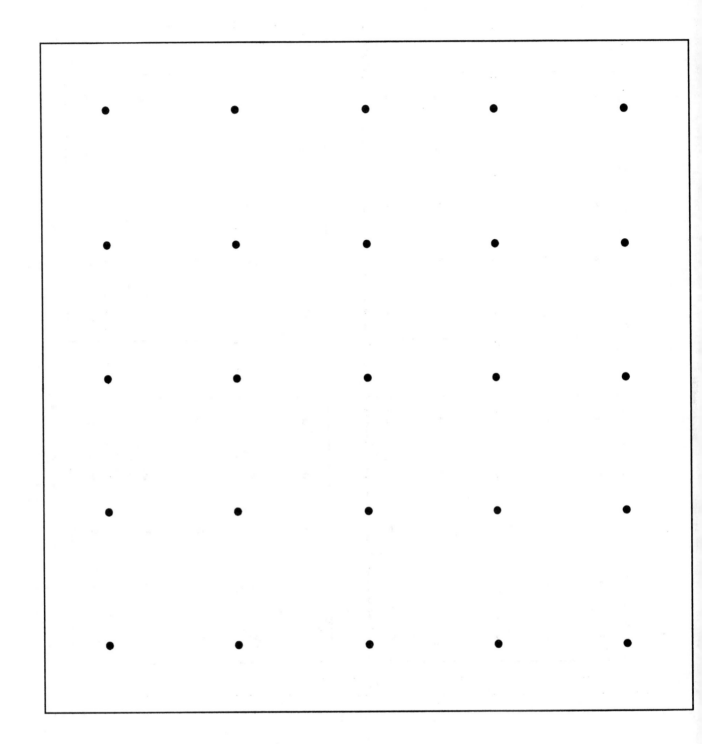

Geoboard pattern - 45

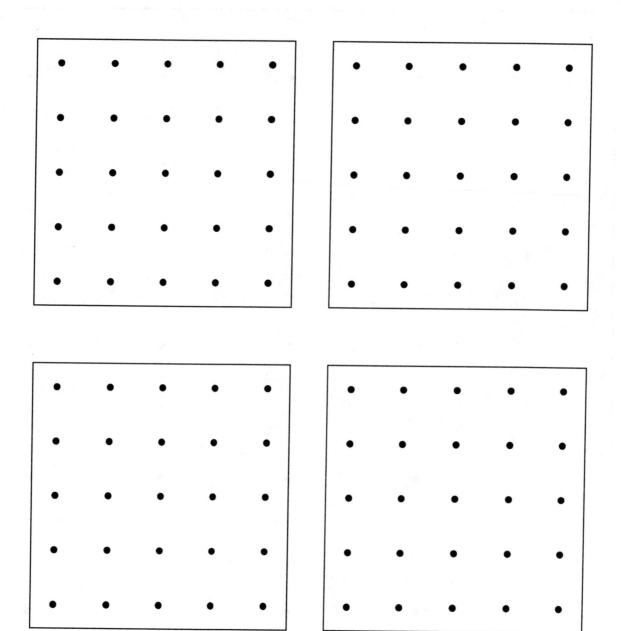

Geoboard recording sheets—46

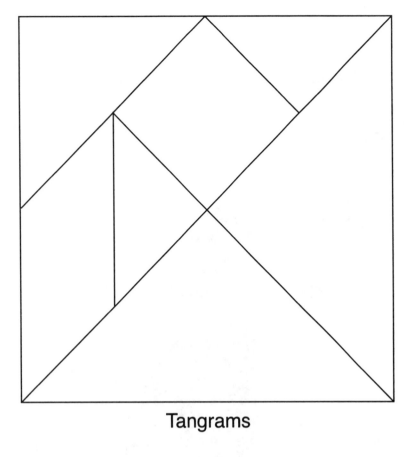

Tangrams

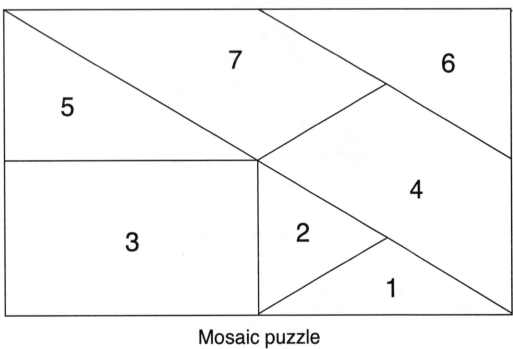

Mosaic puzzle

Tangrams and Mosaic Puzzle - 47

Motion man—Side 1

Directions:

Make copies of Side 1. Then copy Side 2 on the reverse of Side 1. Check the orientation with one copy. When done correctly the two sides will match up when held to the light.

Motion Man—48

Motion man—Side 2
(See directions on Side 1.)

Motion Man—49

Parallelograms

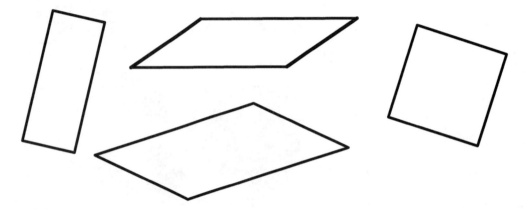

Properties of sides:

Properties of angles:

Properties of diagonals:
  Note: Diagonals are perpendicular or not
             Bisected by the other or not
             Congruent or not

Properties of symmetry (line and point):

Property lists for quadrilaterals—50

Rhombuses

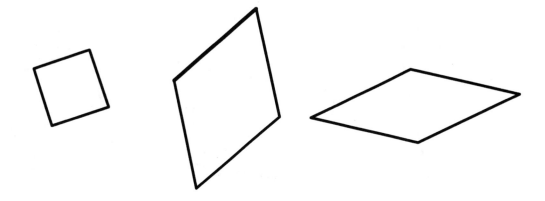

Properties of sides:

Properties of angles:

Properties of diagonals:
Note: Diagonals are perpendicular or not
Bisected by the other or not
Congruent or not

Properties of symmetry (line and point):

Property lists for quadrilaterals—51

Rectangles

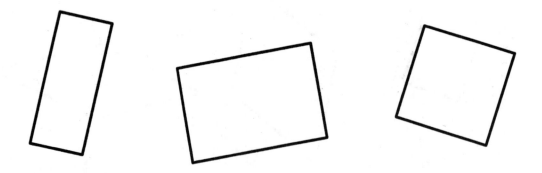

Properties of sides:

Properties of angles:

Properties of diagonals:
      Note: Diagonals are perpendicular or not
                Bisected by the other or not
                Congruent or not

Properties of symmetry (line and point):

Property lists for quadrilaterals—52

Squares

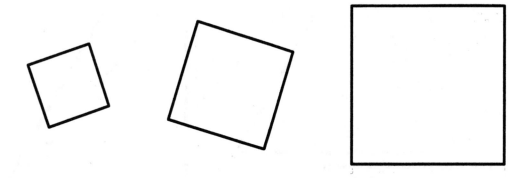

Properties of sides:

Properties of angles:

Properties of diagonals:
   Note: Diagonals are perpendicular or not
              Bisected by the other or not
              Congruent or not

Properties of symmetry (line and point):

Property lists for quadrilaterals—53

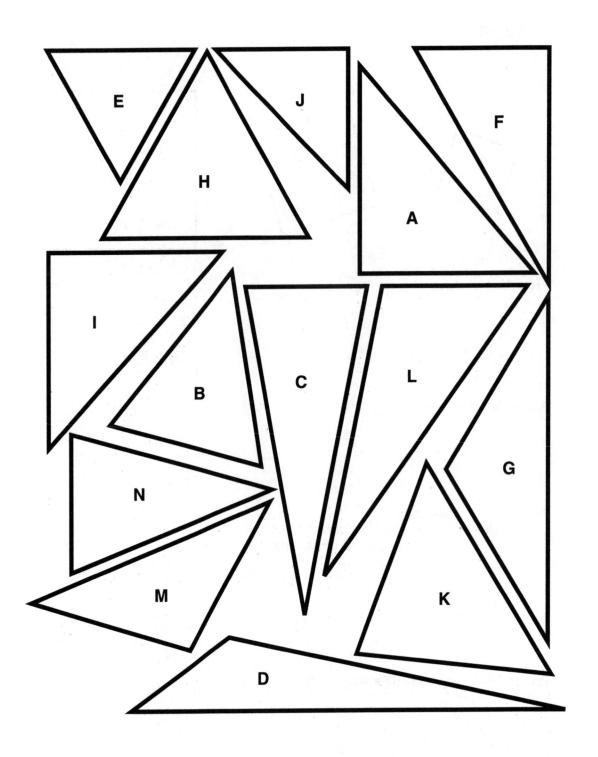

Assorted triangles—54

Woozle Cards—55

NOTES

NOTES

NOTES

NOTES